All Sta

A JOURNEY THROUGH 150 YEARS

An exhibition from
the Centre de Création Industrielle,
Centre Georges Pompidou, Paris,
initiated and directed by Jean Dethier

SCIENCE MUSEUM LONDON

Translated from the French *Le Temps des Gares*

First published in France in 1978 as *Le Temps des Gares* by the Centre Georges Pompidou, Paris
Published in Germany in 1980 as *Die Welt der Bahnhöfe* by Elefanten Press Verlag, West Berlin
Published in Spain in 1980 as *El Mundo de las Estaciones* by Ministerio de la Cultura, Madrid

ISBN 0 901805 16 5

Colour section printed in France by L'Imprimerie Lang, Paris
Filmset and printed in Great Britain by BAS Printers Limited, Over Wallop, Hampshire
Bound in Great Britain by Western Book Co. Ltd., Maesteg, Glamorgan

This exhibition was conceived and assembled by the Centre de Création Industrielle, a department of the Centre national d'art and de culture Georges Pompidou, Paris. The exhibition was produced with the collaboration of foreign institutes:

in Belgium: the Exhibitions Society of the Palais des Beaux-Arts, Brussels
in Britain: the Science Museum, London, and its outstation, the National Railway Museum, York
in Holland: the Museum of Architecture Foundation, and the Documentation Centre for Modern Architecture, Amsterdam
in Italy: the Leonardo da Vinci Museum of Science and Technology, Milan, and the City of Milan

After its presentation in Paris, from 13 December 1978 to 9 April 1979, the exhibition toured France and Europe. It was shown at the Science Museum, London, from 21 May to 27 September 1981.

Associates in the preparation of the exhibition

Belgium
Director: Michel Baudson
President: Karel Gheirlandt
Research: Michel Louis
The Exhibitions Society of the Palais des Beaux-Arts, Brussels
Archives d'Architecture Modern, Brussels

Great Britain
Director: Dame Margaret Weston, Science Museum, London
Keepers: John van Riemsdijk, Science Museum, London; John Coiley, National Railway Museum, York

Holland
Director: Fons Asselbergs
Research: Frank den Oudsten
Museum of Architecture Foundation, Amsterdam

Italy
Director: Pierpacio Saporito
President: Francesco Ogliari
Leonardo da Vinci Museum of Science and Technology, Milan

Exhibition Production Team (Centre Georges Pompidou)

Exhibition director   Jean Dethier
Production assistants   Lydia Elhadad, Lise Grenier, Christine Bancou, Claire Scemia
Architectural models   Alain Pras
Picture research   Sabine Berthier, Arielle Rousselle
Photographers   François-Xavier Bouchart, Jean-Claude Planchet
Artists or illustrators invited to create an original work for the exhibition
Dominique Appia, Daniel Authouart, Michèle Blondel, Michel Dubré, Peter Klasen, Alain Kleinman, Jean-Claude Latil, Fabio Rieti, Denis Rivière, Jean-Claud Silberman
Artistic coordination   Jana Claverie, Mireille Sueur
Audio-visual montage   Patrick Arnoud
Exhibition itinerary   Nicole Richy

For this English edition of the catalogue, the Centre Georges Pompidou would like to express especial thanks to those individuals and institutions in Great Britain who helped with the preparation of the original exhibition and its presentation in London, as well as in the adaptation of the catalogue:

Pippa Bignell; Marcus Binney; British Rail; John Coiley; Gillian Darley; Ruth Eaton; John Harris; the staff of the National Railway Museum, York; Andreas Papadakis; RIBA, drawings collection; SAVE; the staff of the Science Museum; SNCF; Denis Sharp; Lindsay Sharp; Jack Simmons; John van Riemsdijk

This exhibition has been presented under the following titles in different cities in Europe:

1978	'Le temps des gares'	Paris	Centre Georges Pompidou
1979	'Le temps des gares'	Lyons	Espace Lyonnais d'Art Contemporain
1979	'Il mondo delle stazioni'	Milan	Museo Nazionale della Scienza e della Tecnica
1980	'Le temps des gares'	Brussels	Palais des Beaux Arts
1980	'De wereld van het station'	Delft	Technische Hogeschool
1980	'Die Welt der Bahnhöfe'	Berlin	Staatlische Kunsthalle
1980	'El mundo de las estaciones'	Madrid	Palacio de Velasquez del Retiro
1981	'El mundo de las estaciones'	Barcelona	Mercat del Born
1981	'All Stations'	London	Science Museum

Contents

The station: a modern-day Tower of Babel

Railway stations . . .? They are at one and the same time 'volcanos of life' (Malevitch), 'the most beautiful churches in the world' (Cendrars), 'palaces of modern industry where the religion of the [19th] century is displayed, that of the railways. These cathedrals of the new humanity are the meeting points of nations, the centre where all converges, the nucleus of huge stars whose iron rays stretch out to the ends of the earth' (Théophile Gautier).

For a century and a half, stations have been the pivots, the places of command and articulation of a railway empire the spread of which has profoundly marked the structure of a very large number of countries; it has transformed our environment and thus our relationship with the natural, social and cultural surroundings; through speed and reduction in distances, there exists henceforth a new rapport with space and time. Nevertheless the railway station is so ingrained in the very network of our daily routines that one no longer notices it, one no longer sees it: generally one does no more than suffer it.

If we have chosen to consider the station over and above the railway, this is because it is to the railway what the heart is to the vessels in the circulation system. It forms its muscles and regulates its flow: it is the organ of arrival and departure.

The railway station is one of the rare public buildings produced by the industrial revolution which illustrates admirably, over a hundred and fifty years, the gropings, fluctuations and transformations of our Western society. Stations reveal the myths and realities of the epic times we live in. A veritable microcosm of industrial society, a public place where all social classes rub shoulders, the station has been throughout its history at the heart of the present, the many-faceted mirror of a striking array of achievements.

Anchored to the foundations of our industrial system, built on the initial principle of conquest of territory, of business and profit, erected on the mythical ideal of communication of goods and people, and of the peaceful unification of nations, the station is a modern-day Tower of Babel. This exhibition proposes to explore the foundations and many levels — old and recent, real and imaginary — of this tower, which is both ruin and building-site, both familiar and mis-understood: it attempts to illuminate the fragments of which the visitor would be the archaeologist and the futurologist, in other words the interpreter.

There are innumerable ways of assessing railway stations, and that in itself is proof of their remarkable power of suggestion. To cover them more closely, we have chosen to consider them from different aspects: architecture, urbanism, technology, decorum, art, popular culture, politics, strategy, order, discipline, the poetic and the imaginary. Through these comple-mentary angles we hope to provoke a new look at stations and thus at our daily environment. In examining this setting of progress with which the station is branded, our purpose is not to maintain a preoccupation with the past any more than a smug confidence in the future, but to confront as objectively as possible the effects induced by a system existing for a century and a half. This is to allow us better to appreciate its evolutions, better to compare the various aspects both of its history, which has already sometimes been submerged in myth and fantasy, and of its present, in turn exciting and depressing.

The railway station is not a simple place. It expresses forcefully the multiple paradoxes of our society, and brings together certain contradictory ideas which characterize industrial society. The great 19th-century stations are formed of two fundamental elements: the 'passenger building', the construction of which was entrusted to architects — most of whom were resolutely attached to past styles — and the great metal 'shed' covering the platforms. The latter was the responsibility of engineers, conceived by them as a new constructive system and put up with faith and optimism for the technological progress still to come. Nevertheless the smooth functioning of a station depends on the complement of these two elements. So the two extremes of language and ethics in the act of construction confront each other in one and the same place. The station seems, in all aspects, like the simultaneous and contradictory expression of the wonder and tragedy of modern times.

For a century the station had been a place symbolizing progress and limits to surpass: the trains attempting faster and faster speeds rivalled the engineers in the stations who put up more and more audacious superstructures above the platforms, frameworks whose enormity — sometimes bordering on the impossible — seem to want to contain and master the microcosm of society that the station represents, in one and the same mythical projection. So the megalomania of modern times, the cult of technological performance which was to guide the destiny of our civilization and feed a phantasmagoria of a new spirit, displays itself in large city stations. The station becomes a temple to technology wherein is expressed the ritual of a new cult. The spectacle of a steam engine entering the station remains, well after the disappearance of such locomotives, a synthetic vision of promising splendours and of the majesty of industrial civilization; this image will capture the imagination of future generations, who likewise will project their fantasies on it.

If the work of engineers in the conception of 19th-century stations expresses the confidence of a minority in a technological and productivist future, on the other hand the proposals of the architects affirm a contrary feeling reflecting

the majority of public opinion: fear of too sudden a jump into the future, desire of a very cautious dose of tradition and innovation. So in order to disguise the upheavals of the introduction of the railway into the town, the quasi-totalities of 19th-century station buildings take on the appearances of Greek temples and Roman baths, Romanesque basilicas and Gothic cathedrals, Renaissance châteaux and Baroque abbeys. This remarkable continuity in resorting to pastiche and historical fetishism represents a real fear of the coming of a modernity which, even then, worried more than it reassured. This uneasiness at the sight of the station, the alarm provoked by this ultimate place of movement, is felt even today by a good many of our contemporaries. Railway architecture therefore attempted for a century to sublimate this fear of a modernity experienced as an aggression. The projects conceived by those holding the reins of contemporary architecture which were imbued with an epic breath of modernity were almost never realized: of the outbursts of Futurist or Expressionist architects at the beginning of the century, there remain only some plans of stations obstructed in the blind alleys of history.

However, with the Twenties, architecture was to express the definitive victory of the supporters of an international model of industrial society adhering to an ideal of productivity. From then on, architects started to express accurately this stance in inventing a new ethic and a new language of 'international style', the neutrality of which voluntarily affirms nothing other than a machine cult, an incessant preoccupation with a constructive rationality and an operational functionalism. For half a century the vast majority of modern stations have embodied this dominant ideology. It represents the characteristics of our society: an overwhelming uniformity, an indifference to its surroundings and

to the public. It represents the cold rationality which leads our planners to decide everything which composes our environment, with the greatest contempt for cultural, symbolic or emotional aspects, the predominance of the quantitative over the qualitative, the rejection of peculiarities appealing to our senses. The station has become a neutral place whose emptiness frightens even the technocrats, who now try to offset the malaise by broadcasting supposedly comforting, innocuous music. All over the world, new stations have almost abandoned the exterior signs of their civic vocation, the architectural structure of a forum of public life. They are limited to copying dominant examples of the current economic system — the shopping centre or the office building — through which appears the profit incentive that openly seems to exclude any other inclination.

Through the familiar excesses of a rigid functionalism, without cultural references the public expects, modern stations are so empty of feeling that, as compensation perhaps, those of the 19th century now appear generous and eloquent, like the expression of a fantastic and engaging folly. A folly of architecture and ornament, whose very theatrical display in its space had engendered a sort of modern fairyland. This extravagance expressed itself all the more exuberantly as railway imperialism developed in every part of the world where Western industrial power had not yet taken root. So the station was a temple to the fundamental achievement of industrial society: a new, more efficient 'space-time' rapport. The station had to materialize this marvel — which feeds a whole contemporary mythology — in a memorable way.

In the 19th century, the rational application of railway timetables led to a standardization of time over large areas in a number of countries. Station time opposed traditional time, celestial and solar, with a pagan and technological substitute. By imposing over the modern city this unified time, deemed official and national, the station has been endowed with a simultaneously symbolic and functional element, impressive and authoritarian: a great tower punctuated with large clocks which now vied with the old landmarks of the industrial city: the church and belfry. In Europe, station towers are most rare in the Latin countries, where the notion of time is less strict, whereas they abound in countries renowned for their social discipline. In the latter, the station remains one of the places where the obsession for punctuality is felt the most strongly, where measured time has a hold over the modern-day city.

By the frantic pace of the regular rhythms which industrial society has imposed on the world, the station engulfs and disgorges its millions of commuters as a daily implacable ritual; millions of sterile hours are used up in this shooting between work and family, this human flow, this debasement into a long daily routine, between suburban and city stations.

This daily drama, this enormous waste of human energy, is related to the inability of industrial society ever since the 19th century to conceive regional planning patterns which would have avoided the desertion of the country for the towns on an enormous scale. Instead of serving as a vector of decentralization, the railway network has often been used as an instrument of political, economic and social centralization.

The crowds of country people who converged on the rural station in the 19th century were uprooted from their soil by the new productivist logic. They were transplanted into the chaos of suburbs and factories which sprang up around the stations of the big industrial cities.

The confusion and alarm of 19th-century emigrants and likewise today's immigrant workers shows in the stations. In major labour centres

such as Zürich or Munich, exploited by a consumer society which is greedy for manpower it refuses to integrate, they find today only the station in which to regroup furtively, ethnically withdrawn, to look for a derisory consolation together in their uprooted condition. For them the station is simultaneously a real symbol of their cultural break, and the umbilical cord which psychologically links them to their homeland.

The brutality with which our society produces and rejects its fringe element is expressed by the station; it is the first and last tie in a system which has become almost exclusively urban.

The station seems to be the only public place in the town where society's class system is institutionalized: the buffets and above all the waiting rooms are graded numerically. In one century, democratization has allowed the station to pass from a division of society into four social classes, to a three-tiered and finally a two-tiered system. Even stations in China conform to this implacable logic: the numerical division of classes has been modestly replaced there by a less harsh and more sensory system of 'hard' and 'soft' which refers to the nature of the seats in the various waiting rooms: either velvet-covered or wooden.

The urban railway station also manifests, as part of the new human ebb and flow it causes, the democratization of leisure activity: during the Thirties, working-class families invaded the stations, leaving on paid holidays for resorts hitherto reserved for the privileged classes. Historic photographs of this phenomenon clearly indicate the amazement of those people who, standing on the arrival platform of their new destination, have exceeded the geographical limits of the suburbs around their factories for the first time, by train.

The railway station is one of the last great picture-books offered to the public for consumption. For

a century it has exhibited a fantastic collection of signs, emblems, frescoes and symbols, displaying a whole iconography on its façades and in its halls. By way of this, the ruling powers have expressed their ambitions and the bourgeoisie have manifested the values they emphasize in the new society: colonization of nations, territorial conquest, the high-priority development of industry and commerce with, as corollaries, the glorification of national and military virtues, and those of the family, religion and work. The realization of this profusion of instructive imagery has been entrusted only to submissive and academically inspired artists. The destiny of modern art takes shape out of this accumulation of commissioned works in the railway station. Wasn't the station in effect the perfect public place for an integration of pictorial and artistic creation in the daily life of the population? Its dynamism could have confronted the entire range of social classes whose journeys made the station daily a microcosm of society. Monet and Courbet in France, Jules Destrée in Belgium, militated in favour of generous intervention in railway architecture by the pioneers of a new art. But this perspective of opening up a living art for travellers was obstructed: only academic art has really ever been used. So the station represents the divorce between art and daily life which typifies industrial society.

It took a revolution of volcanic dimensions, that of the Russians, finally to allow avant-garde artists to endow their railway stations during the Twenties, and to use them as a cultural and political forum with their 'agit-prop' events; this, in an enormous country where the railway was assured of an essential role in all fields of communication. 'The painters and writers will take up their pots of paint without delay, and by means of the tools of their trade, they will illuminate and cover with drawings the side, forehead and chest of the towns, the stations and the ever-fleeing herds of wagons' (Mayakovsky).

If the artists whose creations marked out the course of modern painting were unable to leave a trace of their genius in the railway station, by way of revenge the station was their inspiration for a profound renewal of expression; 'our artists must find the poetry of stations as their fathers found that of forests and rivers' (Emile Zola). From the Impressionists to the Futurists, from the Expressionists to the Surrealists, the railway station engendered a series of artistic manifestations which changed and enlarged our vision of the world: 'the first foundations of a great metaphysical aesthetic are to be found in the construction of railway stations' (de Chirico, 1910). Monet's pictorial variations of the Gare St-Lazare in Paris, de Chirico's representations of Italian station squares, Dali's metaphysical considerations of the station at Perpignan, the surrealist visions of marshalling yards or suburban stations by Magritte and Delvaux are so many significant stages of a new display of modern imagination. Thus those who were going to help define the essence of pictorial modernism were inspired by the modernism of the station scene; they also knew how to awake a sensitivity in popular art which has given rise to a remarkable breadth of imagery: they were to stereotype the railway station as a legendary and fabulous place.

The media were to contribute equally in conveying and amplifying a vision of the Wonderful and the Tragic in the railway station. From its origins, with Louis Lumière's 'Entry of a train into the station at La Ciotat' in 1895, to the most recent films, the cinema has been profoundly marked by this place of theatricality and mobility: the obvious cinematographic application of the station was to launch a new phantasmagoria, a new dimension in modern narrative.

Whether the point of arrival or departure in the adventure, the railway station is also the solemn setting of 'brief encounters' or the place of chance and unexplored sensualities. Various celebrated brothels have tried, with contrivances for noises and vibrations, with decors and scents, to reconstruct in their luxurious rooms a railway atmosphere in which the modern myth of the 'Madone des Sleepings' has taken shape.

The architecture of the 19th-century station often took the physical form of a new city gate; mentally, it was above all our open gate to a fabulous distance, railbound, adventurous and exotic, which has abundantly fed the inspiration of novelists and poets. However, at the beginning of the railways, they were reluctant to extol the station: 'You, modern poet, you detest modern life. You go against your gods, you don't really accept your age. Why do you find a railway station ugly? A station is beautiful' (Emile Zola).

The station has a characteristic importance in modern imagination and in the world of symbols which psycho-analysis acknowledges: it is 'the expression of the unconscious, the point of departure of evolution, of our new material ventures, both physical and spiritual, it is a centre able to evoke the self [. . .]; the station-master represents the director of active, creative and impersonal forces which govern our destiny' (Jean Chevalier).

Even the image of the modern toy is often symbolized by the electric train set studded with passenger stations, shunting yards and freight stations: a system which simultaneously allows the child to project his aspirations in daily adult life and the adult to indulge his desire for power and control over the world. The real euphoria of domination which the railway station toy provides easily explains its popularity with both parties.

The station strongly emphasizes, for the Western powers, the spirit of their imperialist policy. For a century, it upset the fortunes of the world and found in the railway system an authoritative means of internal and external territorial conquest. Bombay railway station in India is the single most colossal monument built in its century in the whole of Asia; its mass dominates the town and glorifies the virtues of the Victorian age and the spirit of eternity of a foreign occupation. In France, the station in Metz faithfully represents the wishes of the Germans, after the annexing of Alsace-Lorraine, to stamp their own German mark on this conquered territory: the station was preventively adorned with effigies of vigilant Teutonic knights in armour. Through its strategic and monumental importance, both affective and symbolic, this station clearly expresses the political logic which caused its existence. The railway station in Milan, built during the Twenties, conveys a disturbing spirit of enormity and imperialist visions, expressed with a megalomaniac delirium of Assyrian-Babylonian inspiration: the architecture serves to support a triumphal and menacing iconography, with a mixture of fascist emblems, references to the grandeur of the Roman empire and glorifications of physical strength and combat.

In the ingenuous spirit of the first railway entrepreneurs, the station was to be one of the living symbols of the meeting, unity and friendship of peoples. However, it has become the place where mobilized men converge on the eve of the cataclysms of war involving modern nations. At the station there unfolded the pitiful scene of victims returning from the front. At the station there ended up a disordered, jagged accumulation of human beings during the civilian exodus, jammed together in a hope of escape. It was at the station that the Nazis started stockpiling human cattle with a view to herding them in the concentration camps organized like some final shunting yard; in camps — as at Treblinka — sometimes camouflaged as phoney holiday resort stations 'to improve psychological efficiency' after the unloading of the survivors of the 'trains of death'.

In the USSR, an undertaking of considerable political, strategic and economic importance has taken shape recently with the Kremlin's decision to construct a new Trans-Siberian railway. The old line is now considered very vulnerable due to its proximity to the Chinese border. All along this new line across the continent, a number of new stations will become pivots of a series of urban developments and giant industrial combines, the poles of one of the biggest operations of the century in terms of a pioneering colonization of a territory. An epic comparable to railway construction in the American West or in the colonies; but here the epic is on the scale of post-industrial society's ambitions, where a sophisticated economic and strategic plan is crystallized around an immense chain of station sites. Thus the USSR is using the railway to open up a new industrial empire in Siberia, perhaps to be one of the mightiest by the end of the century.

In so far as they are the poles of a railway network, stations have constituted one of the major opportunities of contemporary history to invent new concepts in limiting the growth of towns, by creating coherent and autonomous urban entities around new or satellite stations. Industrial society opposed projects of this nature with an almost unanimous refusal when they were proposed in the 19th and early 20th centuries; it thus made the fatal and historic choice of favouring an accumulation of speculative and profiteering short-term operations, which were the direct cause of the segregated and chaotic structure of the immense modern town extensions and suburbs thus created. The misuse made of the railway station's potentialities is remarkable: it could have become the pivot of urbanization and the germ of a new

town-planning, the catalyst of a harmonious decentralized society in new areas; but those who decide have made it the centre of a real estate and building speculation which boosts the interests of the minority. For more than a century events have shown how speculative interests crystallize around the station in the form of real estate promotion, more and more arrogant with regard to its social context, more and more brutal in its urban context. This now sometimes approaches caricature – from Toronto to Utrecht, from Brussels to Nancy, from London to Paris. The railway station has continued to express around itself the destiny of the modern towns, invested with the power and the logic of capital: by its nature, the station offered the town the privilege of a constantly renewed human market-place. Exploitation of its ebb and flow guaranteed profitability.

One of the last great opportunities to restructure the dying inner cities is now being realized behind station buildings, on the immense areas of railway tracks wedged into the town. This considerable urban and political undertaking is well illustrated in Paris by the exceptional breadth of the project labelled 'Seine Sud-Est', focused around the railway property of the Austerlitz and Lyon stations.

With the energy crisis and the rise of ecological consciousness, the railway is returning to the heart of current debates on the politics of transport, this fundamental essential of our economy of exchange. In France, receipts from passenger main line traffic increased by more than 4 per cent between 1977 and 1978. In the USA the government is presently trying, with Amtrak, to reconstruct a national passenger rail network after having let the railways decline and rot for several decades. America used to have one of the most powerfully structured railway systems in the world, but it was sacrificed to the interests of the automobile and petrol industries.

The rivalry between rail and road transport illustrates one of the great struggles of our time: that of two systems which support divergent behaviour and interests.

In Britain services have been improved substantially over the years. Overall, intercity passengers have been increasing by about 3 per cent per annum since 1976 and on routes where High Speed Trains operate the increase is as much as 30 per cent.

At the heart of this important debate, the station remains a place where there rest the options which model our environment and our daily life. It is a real seismograph of the vibrations and convulsions of our society, its dynamism or its decline, its slightest fluctuations. The media often use photographs of deserted stations to illustrate the effects of a national strike, to show the breakdown of the social and economic life of a country. The closure of rural lines and stations in many Western countries contributes to the progressive decline of our countryside.

Thus several faces of this modern Tower of Babel which is the railway station are already in ruins. However, various new projects are simultaneously under construction at the top of it: work is in progress on new railway lines, buoyed up by new stations; these are part of a railway dynamism which for 150 years has continued to bear witness both to the astonishing ability of technological renewal of a still vital public service, and to the confidence of millions of railway employees in the collective field of modern communications.

How can we stay indifferent to the paradoxes of this future, where destiny and progress meet, where the confusion of our society's impetus is summed up in this modern-day Babel?

Jean Dethier
Exhibition Director
(Translated by Richard Foxcroft)

Note: Because of production exigencies, the section of illustrations in colour on pages I–XL is that of the original catalogue prepared for the exhibition (Le Temps des Gares) at the Centre Georges Pompidou. English translations of the picture captions for these pages follow below.

page I

1
Grand Hall of the old Euston Station in London, completed in 1848 by architect Philip Charles Hardwick; destroyed in 1961 to make space for the new Euston. (Coll. RIBA, London. Photo NRM)
2
Self-portrait of a photographer in a station. (Photo François-Xavier Bouchart)
3
A different poetry: the dominion of lights in a station at night. (Photo Ronzel, SNCF)

pages II–III
In the 19th century, the railway station became a new gateway to the city, an imposing shape which stamped its mark upon the landscape, the hub around which great building operations revolved, the bud of new city growth.

1
Portico at the entrance of Euston Station in London, 1837; architect, Philip Charles Hardwick. (Photo NRM)
2
St Pancras Station and hotel seen from Pentonville Road; John O'Connor, oil on canvas, 1884. (London Museum, photo NRM)
3
The station in Toronto, Canada; at right the lower part of the Canadian National tower, originally conceived as the focus of an enormous complex of buildings to be constructed and occupied by the railway. (Photo R. Van der Hilst)
4
The neighbourhood of the new Gare Montparnasse in Paris: a violent break in scale between the traditional *quartiers* and this gigantic complex of buildings which crystallized around the station during the 1960s and '70s. (Photo Interphotothèque, Paris)
5
St-Jean Station in Bordeaux. Constructed in the 19th century on the outskirts of the city, the railway installations now represent an important break between the urban centre (at left) and its suburbs. (Photo A. Perceval)
6
The railway town of New Swindon, built alongside the Great Western Railway in England. At left, the railway workshops; at right, the workers' town; between the two, the church: in the background, the station. Watercolour by Edward Snell, 1849. (Photo NRM)
7
Project by the Italian Futurist architect Virgilio Marchi in 1919 for a city centre arranged in tiers overlooking the station. (Photo Planchet, CCI)
8
Plan of Bedford Park garden suburb, west of London. The myth of the 'country town' structured around a station. (Photo Hounslow Library, Chiswick)

pages IV–V

1
Air Raid on Willesden Marshalling Yard, painting by Norman Wilkinson. Night bombardment of a London goods yard in 1940: the destruction of the railway network by the enemy was a basic military tactic. Stations were regarded as key railway targets. (Photo NRM)
2
Saying goodbye to departing soldiers: Danish illustration of a classic theme, c. 1864. (Photo DSB)
3
Women Railway Porters in Wartime, 1941; painting by William Roberts. In this century, the appearance of stations changed in wartime with the influx of women workers. Mobilized for the war effort, as in other sectors of industry, women replaced men in administration, maintenance and factory work. (Imperial War Museum. Photo NRM)

Early stations were the pride of 19th-century capitalism. They had the specific function of prestige for the companies that built them. As major public buildings, stations are key monuments to the controlling forces of society. Expressed in their design and decoration are powerful political assumptions and messages.

4
Victoria Station, Bombay, architect F. W. Stevens. Constructed between 1894 and 1896, and reputed to be the largest edifice in Asia at that time. Ostentatious symbol of an expanding society which exported European architectural styles whilst trying to incorporate local traditions. (Photo NRM)
5
Doornfontein Station on the outskirts of Johannesburg. Class divisions are reinforced in South Africa by racial segregation, with separate access to trains. (Photo Dahlström, VDR)
6
Gigantic portrait of Mao Tse-Tung on the façade of Peking Station, 1958. Built by the Soviets, before the break between the two countries, in a purported national Chinese style. (Photo Whitehouse, VDR)
7
Stained glass in the *Salon d'honneur* of Metz Station, built 1905–08, architect Kröger. Constructed during the German occupation of Alsace-Lorraine, its decoration evokes imperial power through the image of Charlemagne. (Photo Planchet, CCI)
8
The statue of Lenin in front of the New Finland Station, Leningrad, 1960. Lenin's arm is raised, as if to indicate the path of Socialism in old Petrograd, the first seat of the Bolshevik government. The city was rebaptized with the name of the leader upon his death. (Photo VDR)

pages VI–VII

1
Churchgate Station, Bombay; arrival of a suburban train. A human tide flowing under the implacable eye of a giant clock which scans the to and fro of its daily users. (Photomontage: VDR)

4
French railway poster, Masseau, 1932: the cult of punctuality on the railways. (Photo Planchet, CCI)

As a place where multitudes rub shoulders, the station becomes an essential element of social control. The frescoes which decorate stations use uplifting themes that apply to the specific regional or national situation.

2
Detail of a mosaic frieze (Labouret, 1926) at Lens, northern France, celebrating the work of the miners of the region. (Photo Bouchart, CCI)

3
Detail of a fresco in the station at Bienne, Switzerland, exalting the family. (Photo Bouchart, CCI)

5
Detail of a frieze in the station of Oporto, Portugal, commemorating great military events. (Photo Evrard and Bastin)

6
Another detail from Oporto, honouring religious virtues (Photo Evrard and Bastin)

In spite of its apparent order, the station is also a place where people living on the periphery of society gather, whether travelling or not. Myth and reality exist side by side and sometimes merge in the characters of literature and cinema.

7
Bombay Central Station: porters lying down on the platforms. (Photo F. Coulon, Atlas Photo)

8
Milan Station, painting by E. Chambon, 1952. (Private coll.)

Stations are organized on military lines. References abound to order and discipline, alluding to railway personnel as well as to travellers.

9
Portrait of the Director of Dutch Railways; painting by J. H. Moesman, 1943. (Utrecht Museum, photo SMU)

10
Figurines of British railway workers, 1860–90, their uniforms inspired by military traditions. (Photo NRM)

pages VIII–IX
Until about 1930, the station often served as a focus for elaborate town settings, adding a theatrical element to the atmosphere of large public spaces. By comparison, recent station designs

show a depressing poverty and illustrate a serious loss of the building's identity in relation to its social and urban context.

1
Arrival of the First Train at Basle Station, by E. Kirchner, oil on canvas, 1884. (Photo Historical Museum of the City of Basle)

2
Pleasure and its . . ., by Jacques Monory, oil on canvas, 1976. The grand monumental staircase built in front of St-Charles Station in Marseilles. (Coll. and photo Galerie Maeght, Paris)

3
Plan for an unidentified train shed, 1883, signed Driver. (Photo NRM)

4
Illustration of a construction set which, to simplify things for the child, reduces the station to two decorative façades. (Coll. C. Wydooge, Heemstede, Holland)

5
The new Gare Montparnasse in Paris, seen from the platforms: a reflection of the worldwide trend in architecture and urban planning in the 1960s, with its aim of increasing the commercial value of land and property.

pages X–XI
The decoration of stations, as major public places, was considered for a century a social and cultural requirement. This sometimes led to lyrical flights of fancy which many now regard with amusement or contempt, for we have become used since the 1930s to an architectural austerity that renounces all ornamentation and any reference to familiar symbols.

1
Mortuary station, built in 1868 at Rookwood, Wales, later dismantled and reconstructed in 1958 in Canberra, Australia. At the station entrance 'Angels of Fame' received travellers on their last journey. (Photo RIBA)

2, 5
A detail and the whole fresco in the hall of Bruges Station, Belgium, which commemorates the main events in the history of the medieval city. (Photo Bouchart, CCI)

3
Large fresco in the booking office of the Gare de Lyon, Paris, c. 1900. A contraction into some 30 metres of the 800 kilometres of landscape along the route of the PLM Railway from Paris to Marseilles via Lyons. (Photo Bouchart, CCI)

4
Mural in the waiting room of the station at Bienne, Switzerland. (Photo Bouchart, CCI)

6
Detail of the ornamentation of columns in the original station in Florence, Italy. (Photo RIBA)

pages XII–XIII
At the same time as they sort out and distribute individuals, stations have from the very start constituted an environment within which all social classes have come together and rubbed shoulders. However, the image preferred by the railway companies favoured their bourgeois clientele. Contemporary imagery both transforms and prolongs this elitist view by depicting the professional man with his attaché-case using the Trans-Europ Express, or in Britain the High Speed Train.

1, 4
Decorated windows in the Gare St-Lazare, Paris, showing the areas served by the network: from the industrial suburb with its landscape of factories – Clichy – to a small seaside resort – Les Sables d'Olonne. (Photos CCI and SNCF)

2
SNCF advertising photo. (Photo Doisneau)

3
Railroad Station, by J. Munk, c. 1850, oil on canvas. (Postal Museum, Frankfurt-am-Main; photo Charmet, Atlas Photo)

5
Poster by Derovet-Lesacq, 1939. A worker heads for the beaches after winning the struggle for paid vacations. (Photo Charmet, Atlas Photo)

6
Waiting at the Station, painting by James Jacques Tissot, 1874. (Dunedin Art Gallery, New Zealand, photo NRM)

7
Railway poster, c. 1920, showing the Great Eastern Hotel at Liverpool Street Station, London. (Photo NRM)

8
Fashion watercolour, 1925. (Photo Charmet, Atlas Photo)

9
SNCF advertising photo – the restaurant in Grenoble Station. (Photo Dewolf, SNCF)

10
SNCF advertising photo for the Trans-Europ Express, 1954. Businessmen – concerned with saving time and therefore money – are the inheritors of a fading tradition, that of the aristocratic traveller. (Photo Lafontant, SNCF)

pages XIV–XV
Although it has almost vanished from the Western landscape, the country station – which forms a link between the land and the city – continues with its crowds of peasants and its market atmosphere, to thrive in the Third World.

'Behind the corner of the station, the peasant women of the neighbouring villages were lined up with their cucumbers, their boiled beef, their clotted cream, their bran loaves. The train stopped, the passengers came out, the crowd mixed together and business was brisk.'
(Boris Pasternak, *Dr Zhivago*, 1958)

1
Richmond Station, Yorkshire. Designed in 1846 by the architect G. T. Andrews, it was sold in 1969 and converted into a garden centre. (Photo NRM)
2
Railway station in Africa. (Photo Bouchart)
3
Railway station in India. (Photo Sée)

The most ordinary and daily use of the station is that of suburban commuters shuttling between their homes and their places of work. The Gare St-Lazare in Paris receives 115 million passengers a year; 95 per cent of them come from the suburbs, and this traffic makes St-Lazare the busiest commuter station in the world.

4
In the Gare St-Lazare, painting by J. Enders, 1900. (Museum of Modern Art, Paris, photo Bulloz)
5, 6
Gare St-Lazare. (Photos Bouchart, CCI and SNCF)
7
Ceramic decorative motif in the station at Pato, Portugal. (Photo Evrard and Bastin)
8, 9
Suburban stations near Paris. (Photos SNCF)

pages XVI–XVII
'Conquest of the world, of distance, of space, of time': for Lamartine in 1838 such was the promise of the railway. These iron rails carried faith in science and progress into the second half of the 19th century. Points of departure and arrival, as well as meeting-places, stations inspired ornamental messages of this optimism, which also sought to reinforce the seductive powers of travel. The images bearing these messages have remained, but the original meaning is lost. The same is true of the mythical and magical aspects of travel, which for a long time nourished writers and film-makers, and now are no more. Gone are the popular pleasure trains which stimulated the humorists, gone are the great European expresses, the sumptuous sleeping cars and Pullmans filled with diplomats and adventurers. They have been replaced by the Trans-Europ Express (TEE) and other trains for businessmen, trains overloaded in vacation periods and peak hours. If it were not for images from the past, the original dream might have faded from our memories . . .

'There is nothing more beautiful about travelling than the railway companies' advertisements in the stations; distant towns are somehow not really desirable except on the iron signs fixed to the side of a train.'

Paul Morand, 1976

1
Buffet of Kazan Station, Moscow. (Photo Garanger, Sipa-Press)
2
The Train of the Husbands, anonymous painting. Seaside resorts, which became popular during the Second Empire in France, enjoyed increasing success. Husbands, kept in town by business and able to visit their pleasant holiday homes only at weekends, travelled to them on trains run specially for their benefit. (Photo Atlas Photo)
3
The Sunday Train, painting by Detti, 1884. (Musée Carnavalet, photo Bulloz)
4
Winter sports station. (Photo VDR)
5
British poster. Parked on the sidings of small rural stations, disused carriages, renamed 'camping coaches', were specially fitted out with eight beds, a dining-room and a kitchen. They were rented to families on vacation. This practice, which appeared in the 1930s, has gradually disappeared since the War. (Photo NRM)

pages XVIII–XIX
Considered the height of modernity and technology, stations have long played an educational role in games and pictures for schoolchildren.

1
Educational print by Delmas, France. (Photo Planchet, CCI)
2
Dutch educational print. (Photo SMU)
3
Boxed game, 'The Little Stationmaster'. (Toy Museum, Poissy, France, photo Planchet, CCI)

Beyond their strictly utilitarian role in transport, stations soon became social spaces with a powerful attraction for multitudes of travellers who found in them a focus for their worldly, cultural or frivolous aspirations.

4
Fashionable display at the opening of the London and Birmingham Railway in 1837; engraving of the scene at Chalk Farm by E. F. Edington. In the 19th century at the launch of each new line, the middle classes seized the opportunity to see and be seen, much as they did at the races or the theatre. (Photo VDR)
5
Royal Station Hotel: British publicity poster. The station hotel became a salon for sophisticated people. (Photo NRM)
6
A steam festival at Shildon Station on 31 August 1975 celebrating the 150th anniversary of the Stockton and Darlington Railway, the first public railway in England to be worked by steam. (Photo Jean Dethier)
7
At the Gare Montparnasse, Paris, a ballet as

part of a new kind of cultural campaign to liven up public places, Christmas 1977. (Photo Y. Patrice)

pages XX–XXI
The Railway Station, painting on canvas by W. P. Frith, 1863. One of the most famous British paintings of the 19th century, it shows Paddington Station, London, designed by Isambard Kingdom Brunel and Sir Matthew Wyatt. (Royal Holloway College, London; photo Picturepoint)

pages XXII–XXIII
From the beginning the station was an interface between the differing methods of architects and engineers. In designing a building for travellers, the former often turned to styles perfected in the past, whereas the latter, in order to create larger and larger edifices over the platforms, sought to extend the capacities of available materials: wood, then wrought iron, cast-iron, steel and concrete. These experiments resulted in a new vocabulary of forms, structures and spaces whose combinations led to constructions of remarkable quality and of unprecedented daring. Architects attempted gradually to assimilate this new ethic of building by renouncing all pastiche and by paying homage to the station's functional truth.

1
Cross-section of La Rochelle Station, France, 1910–23; architect, Pierre Esquié. (Photo Planchet, CCI)
2
The great curved roofs of York Station, England, 1871–77. Thomas Prosser, Benjamin Burleigh and William Peachey were the architect-engineers responsible. (Photo NRM)
3
Elevation of the roof framework of the neo-Gothic station at Bruges, Belgium. (Photo AAM)
4
Train shed of Cologne Station, West Germany, 1889–94. (Photo VDR)
5
Axonometric section of the roof of Lille Station, France, 1889. (Photo AAM)
6
Iron-roofed train shed of Charing Cross Station, London, 1862–64; architect-engineer, Sir John Hawkshaw. (Photo NRM)
7
Façade of the station in Le Havre, France, 1880; architect, Lisch. (Photo Planchet, CCI)
8
Concrete structure of the train shed of Rheims Station, France, 1930–34. (Photo Bouchart, CCI)
9
Wooden roof for the first stations of the London–Brighton line, 1840; architect, David Mocatta. (Photo RIBA)

pages XXIV–XXV
Barely different at first from other industrial buildings, stations during the 19th century exemplified the broader trends in architectural evolution. This began with a return to historical styles according to national tradition and ended up by about the 1800s with a generalized eclecticism. During the first decade of the 20th century a new current became apparent in the use of regional styles born of a strongly-felt need to affirm specific local character.

1
Copenhagen Station, 1847. (Photo DSB)
2
King's Cross Station, London, 1851–52; architect, Lewis Cubitt. (Photo NRM)
3
Gare de l'Est, Paris, 1847–52; architect, François Duquesney. King's Cross and the Gare de l'Est are the first examples of stations in which the building's façade expresses the semicircular shape of the train shed roof behind.
4
Central Station, Amsterdam, 1881–89; architect, P. J. H. Cuypers. (Photo AMA)
5
Temple Meads Station, Bristol, 1865; architect, Sir Matthew Wyatt. (Photo NRM)
6
Leningrad Station, USSR, 1885. (Photo Garanger, Sipa-Press)
7
Toledo Station, Spain, 1865; designed by Don José de Salamanca. (Photo Renfe)
8
Competition design for Milan Station, 1912; architect, Ulisse Stucchini. The station was built between 1923 and 1931 to different designs by the same architect. (Photo FS)
9
St-Gall Station, Switzerland, 1908–13; architect, Alexandre de Senger. (Photo Bouchart, CCI)
10
Project for a station at Jörn, Sweden. (Photo SJ)

pages XXVI–XXVII
Up to the First World War, and even after, an academic style persisted which emanated from the French Ecole des Beaux-Arts at the turn of the century. Nevertheless, a novelty appeared in the exteriors of stations in the form of a clock-tower which became increasingly distinct from the rest of the building. Just before the War the Futurists contributed a surprising contrast with their projects for stations integrated into mega-structures glorifying movement, impermanence and a mechanized society.

1
Competition design for the station at Frankfurt-am-Main, Germany, 1880; architect, Friedrich von Thiersch. (Photo Technische Universität, Berlin)
2
Competition design for Lausanne Station, Switzerland, 1908. (Photo CCF)
3
La Rochelle Station, France, 1910–23; architect, Pierre Esquié. (Document SNCF, photo CCI)
4
Station of the Benedictines at Limoges, France, 1925–29; architect, Gonthier. (Photo Bouchart, CCI)
5
'Plan for an airport and railway station with funicular', 1914; architect, Antonio Sant'Elia. (Illustration Garibaldi Museum, Como; photo City of Milan)
6
'Study for a station', 1913; architect, Antonio Sant'Elia. (Illustration Garibaldi Museum, Como, photo City of Milan)
7
'Structure of a railway space'; 1919, architect, Virgilio Marchi. (Illustration Museo dell'attore, Genoa, photo City of Milan)
8
Baden Station at Basle, 1913; architect, Karl Moser. (Photo Bouchart, CCI)

pages XXVIII–XXIX
After several attempts arising out of the Art Nouveau and Expressionist movements to find an appropriate contemporary style for stations, important efforts were made during the 1920s and '30s to renew the image of stations in the current Art Déco style or with French regional mannerisms.

But the functionalism in architecture which had already appeared by this period really took hold after the Second World War; rapidly becoming the only system of reference throughout the world, it helped to deprive stations of their formal identity.

1
Competition design for the façade of Karlsruhe Station, Germany; architect, Rudolf Bitsan, 1904. (Photo AAM)
2
Project for the new Euston Station, London, 1939. (Photo NRM)
3
Project for a central station in Belgium, 1932; architect, Renaet Braem. (Photo AAM)
4
Project for Piccadilly Station, Manchester, 1965. Study realized by a group of British Rail architects under the direction of W. R. Healey and R. L. Moorcroft. (Photo NRM)
5
Station at Senlis, France, 1922; architect, G. Umbdenstock. (Photo Bouchart, CCI)
6
Study for a station, 1922; architect R. Mallet-Stevens. (Photo AAM)
7
Noyon Station, 1927–29; architects of the Compagnie du Chemin de Fer du Nord, after plans by Urbain Cassan. (Photo Bouchart, CCI)
8
Vanves-Malakoff Station, Paris region, 1936; architect, Jean-Philippot. (Photo Bouchart, CCI)
9
Gare Montparnasse, Paris, 1962–69; architects, Baudoin, Cassan, de Marien, Lopez, Saubot. (Photo Denimal, SNCF)
10
Suburban station near Paris. (Photo Bouchart, CCI)
11
Grigny Central Station, Paris region, 1974. (Photo Bouchart, CCI)

pages XXX–XXXI
During the 1920s and '30s, but principally after the Second World War, the organization of railways was gravely affected by the depopulation of country districts and the emergence of competitive forms of transportation. The triumphal and monumental station gave place to a railway architecture whose neutrality was gradually lost in the anonymity of the city. This profound change was most sharply seen in the demolition of some of the most prestigious buildings ever to express the grandeur of the railway system. To compound the tragedy went the closure of numerous lines and country stations.

1, 2
The Baths of Caracalla, one of the most opulent public buildings in ancient Rome, served as the model in 1906 for the great hall of Pennsylvania Station, New York. In spite of its splendour and fame, it was demolished in 1963 in the face of public opposition, to make way for a skyscraper. (Photos CCI and McGrath)
3, 4, 6
The British are renowned for their pioneer activities in the field of industrial archaeology; certain relics from the great age of railways are exceptions to that rule. Here is one of the oldest stations in the world, Edge Hill in Liverpool, in its present state of decay — a singular contrast to its former splendour, as shown in old prints. This station, conceived by the famous engineer, George Stephenson, was inaugurated on 15 September 1830. (Photos Bouchart, CCI, and NRM)
5
In the country, though stations are not always demolished, it is the rails which have been removed. Here is the station of Pierrefonds, France, built in 1883 and added to the supplementary inventory of historic monuments in 1977. (Photo Bouchart, CCI)
7
Right in the heart of Paris, an abandoned

part of the railway heritage: the station at Boulainvilliers. (Photo Bouchart, CCI)

8
One example among many in the industrialized countries: in a little town, an abandoned station invaded by vegetation. This is the station at Tynemouth, England, built in 1882. (Photo Bouchart, CCI)

pages XXXII–XXXIII
The 1970s were a period of transition: the architecture of old stations was still unfortunately being disfigured by barbarous modernization; but, in the Third World as well as in Europe, railway stations increasingly became the objects of careful attention to preserve their character and integrity.

1
Example of a small French station whose architecture was recently disfigured by refittings painful to the eye. (Photo Bouchart, CCI)

2
At the same time as the Gare du Nord was being classified as an 'historic monument', this car park was attached to it in the most brutal manner possible. (Photo Diot, SNCF)

3
The visual contrast between the modernism of rolling stock and the historical character of old stations is most striking. (Photo Planchet, CCI)

4
The station at Hendaye-Plage, France: an example of the perfect maintenance of the railway station heritage. (Photo Bouchart, CCI)

5, 7
Stations in Malaysia and Thailand: Kuala Lumpur (right) and Hua Hin (left). Stations in the Third World are often, as shown here, the object of careful protection, which is in contrast with the neglect frequently observed in the West. (Photos F. Huguier)

6
Station of Port Erin, Isle of Man, 1903, architect Joseph McArd. A fortunate example of colourful repainting, which restores to the station all the verve of its earlier days. (Photo J. Coiley)

8
The demolition of the historic station in Zürich was planned for many years in order to make way for a large office building. The saving of this station (and, as a result, of the urban centre) was wholly due to a sudden craze for 19th-century architecture. The illustration shows the station being restored in 1978. (Photo Bouchart, CCI)

pages XXXIV–XXXV
Only rarely do recent stations still express, like these, an attempt at an architectural quality and structure appropriate to their public use. An encouraging trend is the conversion of abandoned stations; with imagination, they can play a vital new role in the community.

1
Great hall of the station at Evry-Courcouronnes at the heart of a new town in the Paris region. One of the most recent examples of French railway architecture, it expresses a will to create an architectural event and an authentic public area. (Photo Mazda)

2, 3, 4
These Stockholm Underground stations are among the very rare recent examples of railway architecture where the designers have gone beyond the hitherto sacrosanct limits of functionalism in order to give real importance to the creation of a poetic and imaginative environment. (Photos Storstockholms Lokaltrafik)

5, 6
Two examples, among many, of disused British stations awaiting an imaginative reuse of their structures. Above, Windsor and Eton Station, 1897. Below, Manchester Central Station, 1880; architect, Sir John Fowler. (Photo Bouchart, CCI)

7, 11
Stations of the 'Chemin de Fer de Provence' at Nice and Digne, France: examples of unused lines threatened with extinction that call for combined reuse and revitalization. (Photos Bouchart, CCI)

8
The main SNCF station in the heart of Nice, on the other hand, has undergone detailed restoration of its façades. (Photo Bouchart, CCI)

9
Project for the conversion of the gigantic station in Cincinnati, Ohio, into a university centre; architects, Hardy, Holzman, Pfeiffer Associates. (Photo Hugh G. Hardy)

10
Station of Porte Dauphine, Paris, partially converted into a restaurant. (Photo Bouchart, CCI)

pages XXXVI–XXXVII
Stations have often been able to accommodate on their walls or to stimulate in the public imagination diverse expressions of popular art.

1, 2
In Portugal many urban and rural stations (here Vilar Formoso and Caldas da Rainha) are profusely decorated with ceramic tiles, called 'azulejos', glorifying the popular traditions of various towns and regions of the country. (Photos Evrard and Bastin)

3
Station scene in ceramic tiles on the façade of a café in the Paris region. (Photo Bouchart, CCI)

4
Faience plates with designs of old and recent stations: above, Willemspoort in Amsterdam (1843), below, Eindhoven (1956). (Photos SMU)

5, 6, 7
German station toys. (Photos Bouchart, CCI)

pages XXXVIII–XXXIX
Stations have inspired in popular art a great variety of imagery, which over the years has made them into legendary and fabulous places.

1
An Epinal print: in the early days of the railways, astonished citizens came to look at the spectacle of the station. (Photo Planchet, CCI)

2
Copper model of a signal gantry at Swindon; the work of a British railway worker. (Railway Museum, Swindon; photo NRM)

3
Banner of the National Union of Railwaymen, portraying the epic struggle between the capitalist wolf and the proletarian. (Photo Snarck)

4
Secrets of a Station Master, gouache by Dominique Appia, 1970. (Photo A. Rey)

5
One of the many Epinal prints on the theme of railway stations — to be cut and pasted together. (Photo Planchet, CCI)

6
German wood-engraving, c. 1840, showing the stations and trains between the towns of Fürth and Nuremberg. (Photo SMU)

7
Transfer-ware milk jug with a picture of one of the first British stations, Edge Hill, built by George Stephenson. (Photo NRM)

page XL

Night view of Metz Station, France. (Photo Bouchart, CCI)

Abbreviations

AAM	Archives d'Architecture Moderne, Brussels
ACL	Archives Centrales des Laboratoires, Brussels
AMA	Architectuur Museum, Amsterdam
BR	British Rail, London
CCI	Centre de Création Industrielle, Centre Georges Pompidou, Paris
CFF	Chemins de Fer Fédéraux Suisses, Berne
CN	Canadian National, Montreal
DB	Deutsche Bundesbahn, Frankfurt
DSB	Danische Staatsbahnen, Copenhagen
FS	Azienda Autonoma delle Ferrovie dello Stato, Rome
IGN	Institut Géographique National, Paris
IGM	Institut Géographique Militaire, Brussels
IRPA	Institut Royal du Patrimoine Artistique, Brussels
NRM	National Railway Museum, York
NS	Nederlandse Spoorwegen, Utrecht
ONST	Office National Suisse du Tourisme, Zurich
RENFE	Red National de los Ferrocarriles Españoles, Madrid
RIBA	Royal Institute of British Architects, London
SJ	Chemins de Fer de l'État suédois, Stockholm
SMU	Spoorweg Museum, Utrecht
SNCB	Société Nationale des Chemins de Fer Belges, Brussels
SNCF	Société Nationale des Chemins de Fer Français, Paris
UIC	Union International des Chemins de Fer, Paris
USIS	United States Information Service, Paris
VDR	La Vie du Rail, Paris

LE TEMPS DES GARES

Dans la ville, dès le XIXᵉ siècle, la gare devient une nouvelle porte de la cité, une masse imposante dont la silhouette domine le quartier, le pivot autour duquel se déploient d'énormes opérations d'aménagement urbain, le germe d'urbanisations nouvelles.

1
Portique d'entrée de la gare de Euston à Londres, 1837, Philip Charles Hardwick, architecte. (Photo NRM)

2
La gare et l'hôtel de Saint Pancras vus depuis Pentonville Road; O'Connor, huile sur toile, 1884, London Museum. (Photo NRM)

3
La gare de Toronto, Canada; à droite la base de la tour du Canadien National, initialement conçue comme pivot d'un énorme complexe immobilier prévu au-dessus des emprises de la gare. (Photo R. Van der Hilst)

4
Le quartier de la nouvelle gare Montparnasse à Paris : une violente rupture d'échelle entre les quartiers traditionnels et cette gigantesque opération immobilière cristallisée autour de la gare au cours des années 60/70. (Photo Interphotothèque, Paris)

5
La gare Saint-Jean à Bordeaux : alors que la gare était construite tangentiellement à la ville au XIXᵉ siècle, les emprises des voies ferrées constituent maintenant une coupure importante entre le centre urbain (à gauche) et ses banlieues. (Photo A. Perceval)

6
La ville ferroviaire de New Swindon créée de toutes pièces en 1849 sur le réseau du Great Western Railway en Grande-Bretagne. A gauche les ateliers ferroviaires, à droite la cité ouvrière; entre les deux l'église; au fond la gare. (Photo NRM)

7
Projet de l'architecte futuriste italien Virgilio Marchi en 1919 pour un centre de ville déployé en gradins le long de sa gare. (Photo Planchet, CCI)

8
Plan de la cité-jardin de Bedford Park à l'ouest de Londres : le mythe de la « ville à la campagne » structuré autour d'une gare de banlieue. (Photo Hounslow Library, Chiswick)

Page précédente :

1
Grand hall de Euston à Londres, Philip Charles Hardwick, architecte. Achevé en 1848, détruit en 1961. (Photo NRM)

2
Autoportrait du photographe dans la gare. (Photo François-Xavier Bouchart)

3
Une autre poésie : l'empire des lumières des gares, la nuit. (Photo Ronzel, SNCF)

1

2

3

4

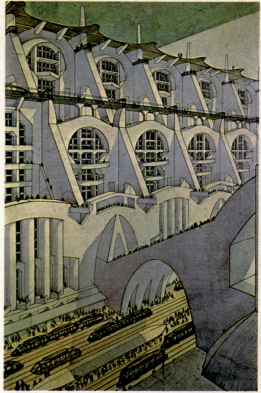

7

5

8

6

1
« Air Raid on Willesden Marshalling yard »
tableau de N. Wilkinson. Bombardement de nuit
d'une gare de triage londonienne en 1940 : la
destruction du réseau ferré aux mains de
l'ennemi constitue un élément de tactique
militaire fondamental; les gares sont
particulièrement visées comme pivots
d'articulation du système. (Photo NRM)
2
Adieux des populations aux soldats mobilisés :
illustration d'un thème classique vers 1864,
Danemark. (Photo DSB)
3
« Femmes-porteurs pendant la guerre », tableau
de W. Robert, Imperial War Museum, Londres.
(Photo NRM)
Dès 1914, les gares verront leur physionomie
se modifier par l'apport massif de la main
d'œuvre féminine. Mobilisées pour l'effort de
guerre, comme dans d'autres secteurs
industriels, elles relayent les hommes
essentiellement à la manutention, l'entretien, la
fabrication de pièces.

**Monument phare du capitalisme, à ses
débuts, avec une fonction première de
prestige, la gare est porteuse d'une
symbolique politique et idéologique.
Théâtre où se joue la représentation des
pouvoirs qui se donnent à voir et impriment
la certitude de leur suprématie dans leur
traduction architecturale et ornementale.**

4
« La gare de Victoria », Bombay, Inde,
F.W. Stevens, architecte. Construite entre 1894
et 1896 et réputée pour être le plus grand
édifice construit en son temps en Asie.
Symbole glorieux d'une société en expansion
qui exporte des styles européens d'architecture
tout en essayant d'intégrer les traditions locales.
(Photo NRM)
5
Gare de Doornfontein dans la banlieue de
Johannesburg. La division en classes se double
en Afrique du Sud des signes de la ségrégation
raciale imposée par le pouvoir blanc; accès
séparés aux trains pour blancs et « non
blancs ». (Photo Dahlström, VDR)
6
Portrait géant de Mao Tsé-Toung sur la facade
de la gare de Pékin, 1958. Édifiée par les
Soviétiques, dans un style qui se voulait
national chinois, dans le cadre de leur
assistance technique, avant la rupture entre les
deux pays. (Photo Whitehouse, VDR)
7
Vitrail du Salon d'honneur de la gare de Metz,
pavillon du Kaiser, 1905-1908, Kröger,
architecte. Édifiée sous l'occupation allemande
de l'Alsace-Lorraine, l'ornementation évoque à
travers la figure de Charlemagne, la volonté de
puissance impériale. (Photo Planchet, CCI)
8
La statue de Lénine devant la « Nouvelle gare
de Finlande » à Léningrad, 1960 : bras levé,
comme indiquant la voie du socialisme dans
cette ancienne Petrograd qui fut le siège du
gouvernement bolchevique et fut baptisée du
nom du son dirigeant à la mort de ce dernier.
(Photo VDR)

1

2

4

7

5

8

6

1
Gare de Bombay-Churchgate, Inde. Arrivée d'un train de banlieue. Marée humaine déferlant sous l'œil implacable d'une horloge géante qui rythme les flux et reflux de ses usagers quotidiens et captifs. (Montage photographique; photos VDR)
4
Affiche pour le réseau français de l'État, Masseau, 1932 : le culte de l'exactitude ferroviaire. (Photo Planchet, CCI)

Lieu de brassage des multitudes, la gare s'impose comme un élément essentiel du contrôle social. Les fresques qui en constituent l'ornementation déploient autant de thèmes de moralisation en accord avec la spécificité régionale ou nationale.

2
Détail de la mosaïque de la gare de Lens (Labouret, 1926) glorifiant le travail des mineurs de fond de la région. (Photo Wieser)
3
Détail de la fresque de la gare de Bienne en Suisse : l'exaltation de la famille. (Photo Bouchart, CCI)
5
Détail de la frise en « azulejos » de la gare de Porto, Portugal, commémorant des hauts faits de guerre; image de la Patrie qui se doit d'être présente. (Photo Evrard et Bastin)
6
Détail des « azulejos » de la gare de Porto : les vertus de la Religion. (Photo Evrard et Bastin)

Mais la gare en dépit de son ordre apparent est aussi un lieu de convergence de marginaux, voyageurs ou non. Mythes et réalités se côtoient et se confondent parfois, au travers des figures qui ont inspiré la littérature et le cinéma.

7
Gare de Bombay Central : porteurs couchés sur les quais. (Photo F. Coulon, Atlas Photo)
8
« La gare de Milan », tableau de E. Chambon, 1952. (Coll. particulière)

L'organisation des gares se constitue sur le modèle militaire. Les références à l'ordre et à la discipline abondent, visant aussi bien les cheminots que les usagers de cet espace.

9
Portrait du Directeur des chemins de fer néerlandais; tableau de J.H. Moesman, 1943, Musée d'Utrecht. Pose austère, horaires à la main sur fond de gare d'Utrecht pour ce symbole de l'autorité. (Photo SMU)
10
Figurines de cheminots britanniques, 1860-1890; uniformes inspirés des traditions militaires. (Photo NRM)

4

7

5

6

8

9

10

Dès ses origines, et jusqu'aux années 1930 environ, la gare a souvent été l'objet d'une mise en scène élaborée dans l'espace de la ville et d'une théâtralisation de ses grands espaces publics. Les réalisations récentes sont, en comparaison, souvent d'une accablante pauvreté et témoignent d'une grave perte d'identité de la gare par rapport à son contexte social et urbain.

1
« L'arrivée du premier train à la gare de Bâle », E. Kirchner, huile sur toile, 1884. (Coll. et photo Musée historique de la ville de Bâle)
2
« Le plaisir et sa... », Jacques Monory, huile sur toile, 1976. Le spectacle du grand escalier monumental édifié devant la gare Saint-Charles à Marseille. (Coll. et photo Galerie Maeght, Paris)
3
Projet de hall de gare non identifié, 1883, signé Driver. (Photo NRM)
4
Illustration d'un jeu de construction qui, pour simplifier le rôle de l'enfant, réduit la gare à deux façades décoratives. (Coll. C. Wijdooge, Heemstede, photo AMA)
5
Les arrières de la nouvelle gare Montparnasse à Paris : un fidèle reflet de l'évolution internationale de l'architecture et de l'urbanisme des années 60 où prévaut l'image de la recherche de rentabilité foncière et immobilière. (Photo SNCF)

1

2

3

4

5

En tant que lieu public, l'ornementation des gares a été considérée durant un siècle comme une évidence première, comme une nécessité sociale et culturelle. Elle a parfois donné lieu à des envols lyriques qui maintenant font souvent sourire ou ricaner tant on nous a habitués depuis les années 1930 à une austérité architecturale reniant toute ornementation et toute référence à un système de signes ou de symboles connus.

1
Gare mortuaire édifiée en 1868 à Rookwood, Pays de Galles, démontée et reconstruite en 1958 à Canberra, Australie. Au fronton de la gare, les « Anges de la Renommée » accueillent l'arrivée des passagers du dernier voyage. (Photo RIBA)

2 et 5
Ensemble et détail de la fresque de la gare de Bruges, Belgique, exaltant les hauts faits de l'histoire de la cité médiévale. (Photos Bouchart, CCI)

3
Hall des départs de la gare de Lyon à Paris. Sur le mur du fond (maintenant en partie caché par de nouveaux guichets) se déploie une immense fresque représentant sans discontinuité les divers paysages traversés par la ligne du PLM, de Paris à Marseille. (Photo Bouchart, CCI)

4
« La ronde des âges », fresque peinte dans la salle d'attente de la gare de Bienne en Suisse. (Photo Bouchart, CCI)

6
Détail de l'ornementation des colonnes de la première gare de Florence, Italie. (Photo RIBA)

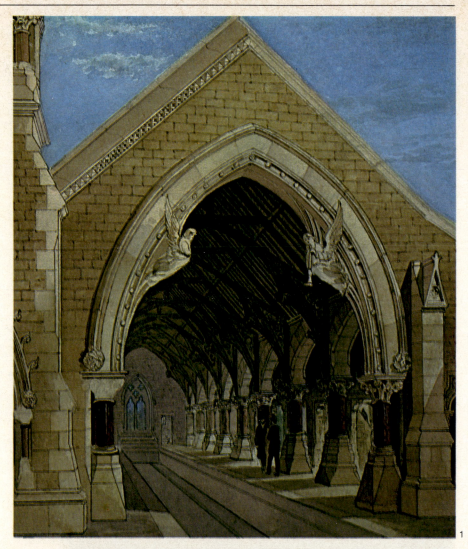

3

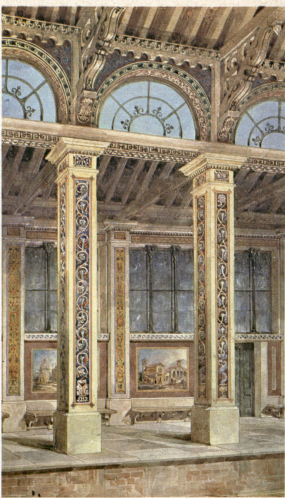

4

5

En même temps qu'elle classe et répartit les individus, la gare constitue, dès ses débuts, un espace où se juxtaposent et se brassent toutes les classes sociales. Cependant, l'image qu'en donnent les compagnies privilégie la clientèle bourgeoise. L'image contemporaine prolonge, en la transformant, cette conception élitiste à travers l'image du cadre à l'attaché-case usager du Trans-Europ-Express.

1 et 4
Vitrages décorés de la gare Saint-Lazare, à Paris représentant les sites desservis par le réseau : de la banlieue industrielle avec son paysage d'usines — Clichy — au petit port de villégiature : Les Sables-d'Olonne. (Photos CCI et SNCF)
2
Document publicitaire de la SNCF. (Photo Doisneau)
3
« Gare de chemin de fer », vers 1850. Toile de J. Munk. (Musée des Postes, Francfort; photo Charmet, Atlas Photo)
5
Affiche par Derovet-Lesacq, 1939. Le prolétaire à l'assaut des plages après la conquête des congés payés. (Photo Charmet, Atlas Photo)
6
« Lady waiting at the station », toile de Jacques Tissot, 1874. (Dunedin Art Gallery, Nouvelle-Zélande; photo NRM)
7
Affiche britannique de 1920 pour l'hôtel de la gare de Liverpool Street à Londres édifié en 1884. (Photo NRM)
8
Aquarelle de mode, 1925. (Photo Charmet, Atlas Photo)
9
Document publicitaire de la SNCF. (Photo Dewolf)
10
Document publicitaire de la SNCF pour le TEE, 1954. (Photo Lafontant, SNCF)

6

8

10

7

9

Alors qu'elle a presque disparu de nos paysages occidentaux, la gare rurale qui faisait le lien entre la campagne et la ville, avec son animation des jours de foire, reste, avec ses foules de paysans et ses ambiances de marchés, caractéristique du Tiers-monde.

« Derrière le coin de la gare [...] les paysannes s'étaient rangées en files, avec leur caillebotte, leur bœuf bouilli, leurs talmouses de seigle... Le train s'arrêtait, les voyageurs arrivaient. Le public s'en mêlait. Le commerce allait bon train ».
B. Pasternak, *Le Docteur Jivago*, 1957.

1
Gare rurale de Richmond Station, Grande-Bretagne. Edifiée en 1846 par l'architecte G.T. Andrews, elle a été récemment reconvertie en « garden center ». (Photo NRM)
2
Gare en Afrique. (Photo Bouchart)
3
Gare en Inde. (Photo Sée)

Une pratique de la gare dans ce qu'elle a de plus banal et quotidien : les fameuses navettes de banlieusards entre lieu de travail et lieu de résidence. La gare Saint-Lazare à Paris, accueille 115 millions de voyageurs par an dont 95 % sont des voyageurs de banlieue : l'importance de son trafic en fait la première gare de banlieue d'Europe.

4
« A la gare Saint-Lazare », tableau de J. Enders, 1900. (Musée d'Art Moderne de la Ville de Paris; photo Bulloz)
5 et 6
Gare Saint-Lazare. (Photos Bouchart, CCI et SNCF)
7
Motif d'une céramique à la gare de Pato, Portugal. Variante poétique de l'emprise du temps. (Photo Evrard et Bastin)
8 et 9
Gares de la banlieue parisienne. (Photos SNCF)

«- Il y a les Bulgares du Nord
*   les Bulgares de l'Est*
*   et les Bulgares de l'Ouest*
*   dits aussi Bulgares Saint-Lazare. »*
Alphonse Allais

« La conquête du monde, des distances, des espaces, des temps », tels sont pour Lamartine les enjeux du ferroviaire en 1838. Voilà les rails porteurs de cette foi dans la Science et le Progrès qui s'épanouit dans la deuxième moitié du xixᵉ siècle. Points de départ et d'arrivée, mais aussi lieux de rencontre, les gares sont investies des messages ornementaux de cet optimisme qui veulent aussi renforcer le pouvoir de séduction des voyages. Ces messages sont restés dans l'imaginaire mais ils ont perdu leur sens initial. Il en va de même du fabuleux et de la magie du voyage. La mythologie qui alimenta longtemps écrivains et cinéastes n'est plus guère qu'un souvenir. Finis les « trains de plaisir » populaires qui stimulaient la verve des humoristes, finis les « grands express européens », les « sleepings » et les « pullmans » somptueux peuplés de diplomates et d'aventuriers. Leur ont succédé les « Trans-Europ-Express » et autres trains d'affaires, les trains bondés des vacances et des périodes de pointe. Sans que, pour autant, le rêve initial ait disparu de nos mémoires...

« Ce qu'il y a de plus beau dans les voyages ce sont les affiches des compagnies de chemin de fer dans les gares, et les villes lointaines ne sont vraiment désirables que sur les plaques de tôles accrochées au flanc des wagons. »
Paul Morand, 1976.

1
Buffet de la gare de Kazan à Moscou. (Photo Garanger, Sipa-Press)
2
« Le train des maris », anonyme. (Photo Atlas Photo)
3
« Le train du dimanche », Detti, 1884. (Musée Carnavalet; photo Bulloz)
4
Gare de sports d'hiver. (Photo VDR)
5
Affiche britannique. Installés aux abord immédiats des petites gares rurales britanniques, des wagons désaffectés, des « camping coaches » étaient spécialement aménagés avec huit lits, une salle à manger et une cuisine. Ils étaient loués aux familles pour la durée des vacances. Cette pratique, apparue dans les années 30, a progressivement disparu après la guerre. (Photo NRM)

1

2

4

3

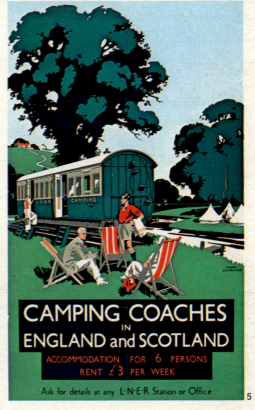

CAMPING COACHES
IN
ENGLAND and SCOTLAND
ACCOMMODATION FOR 6 PERSONS
RENT £3 PER WEEK
Ask for details at any L·N·E·R Station or Office

5

En tant que haut lieu de la modernité, de la technologie, la gare a longtemps eu une vocation pédagogique par le biais des jeux et des planches à l'usage des écoliers.

1
Planche pédagogique Delmas, France. (Photo Planchet, CCI)

2
Planche pédagogique hollandaise. (Photo SMU)

3
Boîte de jeu « Le petit chef de gare ». (Musée du jouet de Poissy, France; photo Planchet, CCI)

Dépassant son rôle strictement utilitaire lié au transport, la gare devient vite un espace social d'un puissant pouvoir d'attraction sur des multitudes d'usagers qui y trouvent réponse à des aspirations mondaines, culturelles ou ludiques.

4
« The London and Birmingham Railway », Primrose Hill, Chalk Farm; gravure de A.F. Edington, 1837-1838. Au début des chemins de fer, la gare est investie comme lieu de promenade élégante des bourgeois, comme lieu d'un nouveau spectacle technologique où il faut se montrer.

5
« Royal Station Hotel » : affiche publicitaire britannique; l'hôtel de la Gare devient un salon de mondanités. (Photo NRM)

6
Fête de la vapeur à la gare de Shildon le 31 août 1975 lors de la célébration du 150e anniversaire de la ligne Stockton-Darlington, Grande-Bretagne. (Photo VDR)

7
A la gare Montparnasse, à Paris, un ballet organisé dans le cadre d'un nouveau genre de campagne culturelle d'animation des lieux publics, Noël 1977. (Photo Y. Patrice)

Pages suivantes :
« The railway station », huile sur toile, W.P. Frith, 1863. Une des toiles les plus fameuses de la peinture anglaise du xixe siècle; elle représente la gare de Paddington à Londres édifiée par I.K. Brunel et M.D. Wyatt. (Royal Holloway College, Londres; photo Picturepoint)

ROYAL STATION HOTEL
YORK
PART OF THE L·N·E·R HOTELS SERVICE

5

6

7

Dès ses origines, la gare apparaît comme un lieu de confrontation abrupte entre les pratiques et les idéologies des architectes et des ingénieurs. Les premiers se réfèrent souvent, pour construire le bâtiment des voyageurs, aux styles révolus du passé, tandis que les autres, pour lancer des halles de plus en plus vastes au-dessus des quais, cherchent à exploiter toutes les ressources technologiques des matériaux disponibles : le bois, puis le fer, la fonte, l'acier et le béton. De ces recherches, il résultera la définition d'un nouveau vocabulaire de formes, de structures et d'espaces dont la combinaison donnera lieu à des réalisations d'une remarquable qualité et d'une audace sans précédent. Les architectes tenteront progressivement d'assimiler cette nouvelle éthique de l'acte de bâtir en renonçant aux pastiches et en exaltant la vérité fonctionnelle de la gare.

1
Coupe transversale de la gare de La Rochelle, France 1910-1923, Pierre Esquié architecte. (Photo Planchet, CCI)
2
La grande charpente courbe de la gare de York, Grande-Bretagne, 1871-1877. Thomas Prosser, Benjamin Burleigh et William Peachey, architectes et ingénieurs. (Photo NRM)
3
Élévation de la charpente de la gare néo-gothique de Bruges, Belgique. (Photo AAM)
4
Halle de la gare de Cologne, Allemagne Fédérale, 1889-1894. (Photo VDR)
5
Coupe axonométrique de la charpente de la gare de Lille, France, 1889. (Photo AAM)
6
Halle métallique de la gare de Charing Cross à Londres, 1862-1864, sir John Hawkshaw, architecte et ingénieur. (Photo NRM)
7
Façade de la gare du Havre, France, 1880, Lisch architecte. (Photo Planchet, CCI)
8
Structure en béton de la halle de la gare de Reims, France, 1930-1934. (Photo Bouchart, CCI)
9
Charpente en bois pour les premières gares de la ligne Londres-Brighton, 1840, David Moccata architecte. (Photo RIBA)

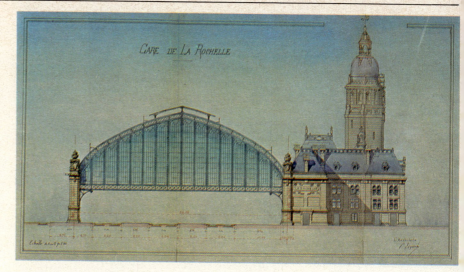

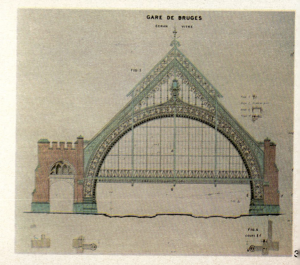

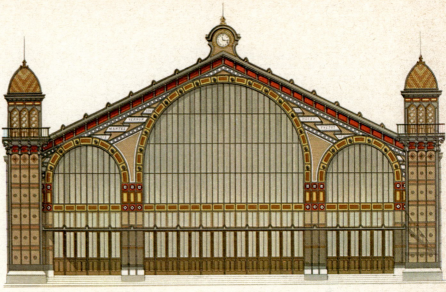

7

4

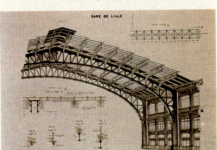

5

8

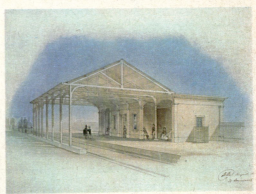

9

6

Peu démarqués à leur origine des autres constructions industrielles, les bâtiments des gares illustreront au cours du XIXe siècle les grandes tendances de l'évolution architecturale par un premier recours à des styles historiques départagés selon des traditions nationales, pour aboutir vers les années 1880 à un éclectisme généralisé. Durant la première décennie du XXe siècle se manifestera un nouveau courant né du besoin impérieux d'affirmer un caractère local spécifique en faisant appel à des styles régionaux.

1
Gare de Copenhague, 1847. (Photo DSB)
2 et 3
Gare de King's Cross à Londres, 1851-1852 Lewis Cubitt, architecte. (Photo NRM)
Gare de l'Est à Paris, 1847-1852; François Duquesney, architecte. Premiers exemples de gares dont le bâtiment exprime en façade, par une ou deux baies, la structure de la halle métallique semi-circulaire construite à l'arrière-plan. (Photo Planchet CCI)
4
Gare centrale d'Amsterdam, 1881-1889; P.J.H. Cuypers, architecte. (Photo AMA)
5
Gare de Temple Meads à Bristol, Grande-Bretagne, 1865, sir Matthew Wyatt, architecte. (Photo NRM)
6
Gare de Léningrad, URSS, 1885. (Photo Garanger, Sipa-Press)
7
Gare de Tolède, Espagne, 1865; Don José de Salamanca, constructeur. (Photo Renfe)
8
Concours pour la gare de Milan, 1912; projet d'Ulisse Stacchini, architecte. La gare sera construite par le même architecte entre 1923 et 1931 selon un parti différent. (Photo FS)
9
Gare de Saint-Gall, Suisse, 1908-1913; Alexandre de Senger, architecte. (Photo Bouchart, CCI)
10
Projet pour la gare de Jörn, Suède. (Photo SJ)

1

2

3

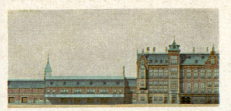

4

5

8

6

9

7

10

Jusqu'à la Première Guerre mondiale — et même au-delà — persistera un style désormais académique issu des recherches de l'École des Beaux-Arts, au tournant du siècle. Une nouveauté apparaît cependant dans l'aspect extérieur de la gare avec la mise en valeur d'une tour de l'horloge qui se détachera de plus en plus du bâtiment. Les futuristes apportent, à la veille de la guerre, un contraste surprenant avec leurs projets de gares intégrées dans des méga-structures glorifiant le mouvement, l'éphémère et la société machiniste.

1
Concours pour la gare de Francfort, Allemagne, 1880; projet de Von Thiersch, architecte. (Photo Technische Universität Berlin)
2
Projet de concours pour la gare de Lausanne, Suisse, 1908. (Photo CFF)
3
Gare de La Rochelle, France, 1910-1923; Pierre Esquié, architecte. (Doc. SNCF, photo CCI)
4
Gare des Bénédictins à Limoges, France, 1925-1929; Gonthier, architecte. (Photo Bouchart, CCI)
5
« Projet pour une gare d'aéroplanes et de trains avec funiculaire », 1914; Antonio Sant'Elia, architecte. (Doc. Museo Garibaldi, Côme; photo Ville de Milan)
6
« Étude pour une gare », 1913; Antonio Sant'Elia, architecte. (Doc. Museo Garibaldi, Côme; photo Ville de Milan)
7
« Structure d'un espace ferroviaire », 1919; étude de Virgilio Marchi, architecte. (Doc. Museo dell'attore, Gênes; photo Ville de Milan)
8
La gare badoise à Bâle, 1913; Karl Moser, architecte. (Photo Bouchart, CCI)

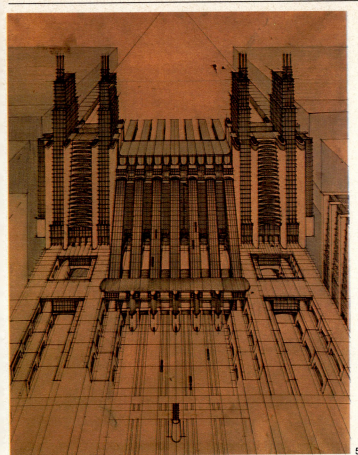

5

7

6

8

Après quelques tentatives liées aux mouvements « Art Nouveau » et expressionniste, on assistera durant les années 20-30 à des efforts importants pour renouveler l'image de la gare avec le courant « Art Déco » ou les recherches françaises, parfois tournées vers le régionalisme.

Mais le fonctionnalisme qui avait déjà fait son apparation à cette époque va connaître un plein essor après la Deuxième Guerre; devenu rapidement l'unique système de références dans le monde entier, il contribuera à priver la gare de son identité formelle.

1
Concours pour la façade de la gare de Karlsruhe, Allemagne; projet de Rudolf Bitsan, architecte, 1904. (Photo AAM)
2
Projet pour la nouvelle gare de Euston à Londres, 1939. (Photo NRM)
3
Projet pour une gare centrale en Belgique, 1932; Renaet Braem, architecte. (Photo AAM)
4
Projet pour la gare de Piccadilly à Manchester, 1965. Étude réalisée par un groupe d'architectes du British Rail sous la direction de W.R. Hedley et R.L. Moorcroft. (Photo NRM)
5
Gare de Senlis, France, 1922; G. Umbdenstock, architecte. (Photo Bouchart, CCI)
6
Étude pour une gare, 1922, par R. Mallet-Stevens, architecte. (Photo AAM)
7
Gare de Noyon, 1927-1929; architectes de la Compagnie du chemin de fer du Nord, d'après les plans d'Urbain Cassan. (Photo Bouchart, CCI)
8
Gare de Vanves-Malakoff, région parisienne, 1936; Jean-Philippot, architecte. (Photo Bouchart, CCI)
9
Gare Montparnasse à Paris, 1962-1969; Baudoin, Cassan, de Marien, Lopez, Saubot, architectes. (Photo Denimal, SNCF)
10
Gare de banlieue parisienne. (Photo Bouchart, CCI)
11
Gare de Grigny-Centre, région parisienne, 1974. (Photo Bouchart, CCI)

5

6

7

8

9

10

11

Dès les années 1920-30, mais principalement après la Deuxième Guerre mondiale, l'organisation du système ferroviaire va être largement affectée par le dépeuplement des campagnes et par l'émergence de modes de transport concurrents : la gare triomphaliste et monumentale fait place à une architecture ferroviaire dont la neutralité va progressivement se noyer dans l'anonymat de la ville. Cette profonde mutation s'exprime dans notre environnement par la démolition de certains bâtiments de voyageurs parmi les plus prestigieux, les plus symboliques de la grandeur du système ferroviaire et par la fermeture de nombreuses lignes et gares rurales.

1 et 2
Les thermes de Caracalla, qui étaient dans la Rome antique un des lieux publics les plus fastueux, ont servi de modèle en 1906 pour la conception du grand hall de la gare de Pennsylvania à New York. Malgré sa splendeur et sa renommée, elle a été démolie en 1963, en dépit de l'opposition du public à ce projet, pour faire place à une opération immobilière. (Photos CCI et Mac Grath)

3, 4 et 6
Les Britanniques sont réputés pour leurs actions de pionniers en matière d'archéologie industrielle; des exceptions notoires dans le domaine ferroviaire confirment cette règle. Ici une des plus anciennes gares du monde, Edge Hill à Liverpool, dans son état actuel de délabrement qui contraste singulièrement avec la splendeur du lieu dont témoignent les gravures anciennes. Cette gare, conçue par le célèbre ingénieur George Stephenson, a été ouverte le 15 septembre 1830. (Photos Bouchart, CCI et NRM).

5
Dans les campagnes, si les architectures des gares n'ont pas toujours été démolies ce sont souvent les voies qui ont été déposées. Ici la gare de Pierrefonds, France, bâtie en 1883 et inscrite à l'inventaire supplémentaire des Monuments historiques en 1977. (Photo Bouchart, CCI)

7
Au cœur même de Paris, un patrimoine ferroviaire à l'abandon : la gare de Boulainvilliers. (Photo Bouchart, CCI)

8
Un exemple parmi tant d'autres dans les pays industrialisés : dans une petite ville, une gare abandonnée envahie de végétation. Ici la gare de Tynemouth édifiée en 1882. (Photo Bouchart, CCI)

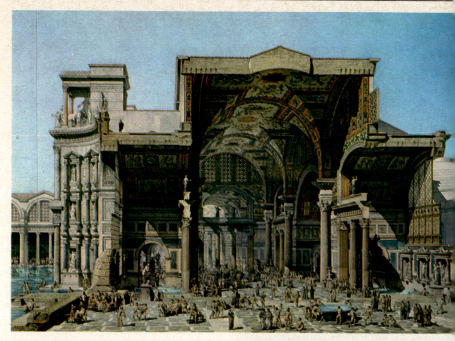

3

6

4

7

5

8

Les années 1970 constituent une période de transition : l'architecture des gares anciennes est encore malencontreusement défigurée par des modernisations brutales mais elle commence, aussi bien dans le Tiers-monde qu'en Europe, à être l'objet de soins attentifs pour préserver son caractère et sa cohérence.

1
Exemple de petite gare française dont l'architecture a été récemment défigurée par des aménagements agressifs pour l'œil. (Photo Bouchart, CCI)
2
Alors que la façade de la gare du Nord à Paris était en cours de classement comme « monument historique », ce parking lui était accolé de la façon la plus brutale qui soit. (Photo Diot, SNCF)
3
Le contraste visuel entre le modernisme du matériel roulant et le caractère désormais historique des gares anciennes est très stimulant. Plutôt que de tenter de neutraliser ces différences par des modernisations dérisoires des bâtiments, il faut chercher à exalter ces particularismes complémentaires. (Photo Planchet CCI)
4
La gare de Hendaye-Plage, France : un exemple de parfait entretien du patrimoine immobilier ferroviaire. (Photo Bouchard, CCI)
5 et 7
Gares de Malaisie et de Thaïlande : Kuala Lumpur, à droite et Hua Hin, à gauche. Les gares rurales ou urbaines du Tiers-monde sont souvent, comme ici, l'objet d'une protection attentive qui contraste avec la négligence fréquemment observée en Occident vis-à-vis du patrimoine historique. (Photos F. Huguier)
6
Gare de Port-Erin, île de Man, Grande-Bretagne, 1903, Joseph Mc. Ard archictecte. Un exemple heureux de polychromie qui restitue à la gare toute la verve de sa jeunesse. (Photo J. Coiley)
8
La démolition de la gare historique de Zurich était programmée depuis de nombreuses années pour faire place à une énorme opération immobilière. On ne doit la sauvegarde de la gare (et, en conséquence, du centre urbain) qu'à un engouement subit de l'opinion pour l'architecture du XIXᵉ siècle. Ici, le chantier de restauration de la gare en 1978. (Photo Bouchart, CCI)

Très rares sont les gares récentes où s'expriment encore, comme ici, une recherche de qualité architecturale et plastique propre à ce lieu public.
De plus en plus fréquentes sont les gares anciennes désaffectées qui se prêteraient bien à des opérations de reconversion du bâtiment permettant, avec de l'imagination, d'y accueillir de nouvelles activités d'intérêt public.

1
Grand hall de la gare d'Evry-Courcouronnes au cœur d'une ville nouvelle de la région parisienne. Un des plus récents exemples d'architecture ferroviaire française où s'exprime une volonté de créer un événement architectural et un véritable espace public. Bernard Hamburger, architecte. (Photo Mazda)
2, 3 et 4
Ces stations du métro de Stockholm constituent un des très rares exemples récents d'architecture ferroviaire où les concepteurs ont dépassé les limites jusqu'ici sacro-saintes du fonctionnalisme pour accorder une réelle importance à la création d'un environnement poétique et imaginaire constellé de références stimulantes à l'histoire et à... l'architecture des grottes. (Photos Storstockholms Lokaltrafik)
5 et 6
Deux exemples, parmi tant d'autres, de gares britanniques désaffectées en attente d'une réutilisation imaginative de leurs bâtiments et de leurs structures. En haut, la gare de Windsor et Eton, 1897. En bas, la Gare centrale de Manchester, 1880, sir John Fowler, architecte. (Photo Bouchart, CCI)
7 et 11
Les gares de tête de la ligne du « Chemin de fer de Provence » à Nice et à Digne, France. Un exemple de desserte ferroviaire menacée de disparition qui appelle une opération combinée de recyclage et de re-vitalisation. (Photos Bouchart, CCI)
8
Au cœur de Nice, la gare SNCF a par contre fait l'objet d'une minutieuse restauration de ses façades. (Photo Bouchart, CCI)
9
Projet de reconversion de la gare géante de Cincinnati (USA) en centre universitaire; Hardy, architecte. (Photo Hardy)
10
Gare de la porte Dauphine à Paris, partiellement reconvertie en restaurant. (Photo Bouchart, CCI)

5

9

6

10

7

11

8

La gare a souvent su accueillir sur ses murs ou susciter dans l'imagination du public des expressions très diverses de l'art populaire.

1 et 2
Au Portugal de nombreuses gares urbaines et rurales (ici Vilar Formoso et Caldas da Rainha) sont abondamment ornementées de décors en céramique, dits « azulejos », glorifiant les traditions populaires des diverses villes ou régions du pays. (Photos Evrard et Bastin)
3
Décor de gare en céramique sur la devanture d'un café de la région parisienne. (Photo Bouchart, CCI)
4
Assiettes en faïence avec représentation de gares anciennes et récentes; en haut, Willemspoort à Amsterdam, 1843, en bas, Eindhoven, 1956. (Photos SMU)
5, 6 et 7
Gares jouets de fabrication allemande. (Photos Bouchart, CCI)

3

5

4

6

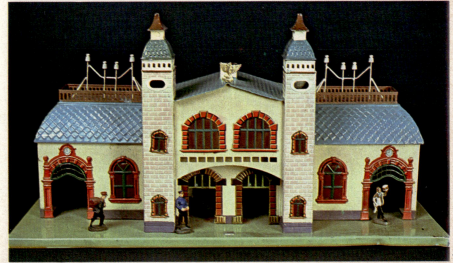

7

La gare a éveillé dans l'art populaire une sensibilité qui a donné lieu à une grande diversité d'imageries : elles vont progressivement la stéréotyper en un lieu légendaire et fabuleux.

1
Imagerie Pellerin, Épinal, France : au début des chemins de fer, les citadins émerveillés venaient assister au spectacle de la gare. (Photo Planchet, CCI)
2
Maquette en cuivre d'un portique de signaux de la gare de Swindon, œuvre d'un cheminot britannique. (Musée ferroviaire de Swindon, photo NRM)
3
Bannière syndicale de l'Union nationale des cheminots britanniques : le combat épique entre le loup capitaliste et le prolétariat. (Photo Snarck)
4
« Confidences d'un chef de gare », Dominique Appia, gouache, 1970. (Photo A. Rey)
5
Détail d'une des nombreuses planches éditées par les Imageries Pellerin d'Épinal sur le thème des gares : à découper et à coller pour en réaliser les maquettes. (Photo Planchet, CCI)
6
Gravure sur bois allemande vers 1840 : les gares et les convois entre les villes de Fürth et de Nuremberg. (Photo SMU)
7
Pot à lait en faïence avec représentation d'une des premières gares britanniques, celle de Egde Hill édifiée par George Stephenson. (Photo NRM)

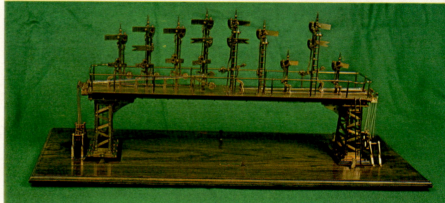

Page suivante :
Vue nocturne de la gare de Metz, France.
(Photo Bouchart, CCI)

4

5

7

6

17

Decor and decorum

Modern stations are meant to be functional and efficient. Their designs are stark in their simplicity. By contrast, the first stations were seen, like other public buildings, as centres of social life and display. The elaboration of those parts of the building associated with the rituals of departure and arrival turned the railway station into theatre, where each traveller knew his role. The station thus became the architectural expression of a collective urban spirit.

Because large numbers of individuals of all origins and social classes passed through it, the station might have become (as shown by Monet and Courbet in France, or Destrée in Belgium) the ideal place to popularize a new art: offering to artists, on the vastness of its walls, a forum of unprecedented scale. But the railway companies and the state preferred to give the job of covering the stations with frescoes and sculptures, allegorizing or edificatory, to academic artists who were already established in decorative painting.

In this way a code of ornament developed that drew its themes from the railway companies, the idea of progress, the destinations served by the line: a sort of hymn of praise for the unification of the country.

But then, in the 1930s, with the competition from road and air transport and with the emergence of an international modern style of architecture, this grandiose quality became outdated, and the station was soon reduced to something neutral with nothing to say either to the eye or to the imagination. The hygienic void which resulted was rapidly taken over by a new set of symbols: advertising began to exploit the presence of the crowds.

Originally conceived somewhat poetically as a space for communal and convivial life, the station has progressively declined into a place programmed for the functions of consumption.

To glorify the wonders of modern
technology which railways
represented, stations were created
with the greatest of care for their
users: façades were treated like stage
sets; booking halls of imperial
grandeur with picturesque decor and
dramatic lighting effects breathed an
air of epic grandeur. Porticoes and
archways of almost operatic pomp
were raised at the entrances to stations
in town and country.

'O miracle!
What a spectacle
Before our eyes!
What prodigy
Exercises his prestige
On these places?
What power
Puts this immense machine
Into motion?
What spirit
Drives and launches
This world and directs it?'
Cantata composed by M. Neumann and
sung on 4 October 1859 during the solemn
arrival of the first train at the Gare de
Luxembourg, Paris.

1
Official opening of one of the first Italian
stations. A theatrical extravaganza to
celebrate the march of technological
progress. (Photo P. Saporito)
2
Opening of the station at Hofors, Sweden,
1858. To mark the opening up of the
countryside by the railway, the local people
transformed a copse of pines into a sort of
baroque architectural fantasy beside the
station. (Photo Swedish Railway Museum)
3
The great hall of Grand Central Station, New
York. A dramatization of public space
designed to produce the most spectacular
luminous effects. (Photo New York Central
Railroad)
4
Façade of the station at Barreiro, Portugal.
(Photo Portuguese Railways)
5
Drawing for the façade of a station planned
by the St-Germain Railway Company at the
Place de la Madeleine, in the heart of Paris.
(Photo CCI)

1

2

3

4

5

As the hub of a railway system which provided industrial society with one of the basic elements in its growth, the station proclaimed its function with lavish pride.

1
The great hall of the North Station in Vienna, 1865. (Musée des Arts Décoratifs, Paris, photo CCI)
2
The great hall of Milan Central Station. Architect, Ulisse Stacchini, 1920–30. (Photo FS)
3
The great hall of the Anhalter Station in Berlin, 1880. (Photo CCI)
4
Entrance hall of Pennsylvania Station, New York, 1906–10. Architects, McKim, Mead and White. (Photo CCI)

1

If the decoration of stations sometimes celebrates that friendship between peoples which – according to its first promoters – the railway would bring, it is more often designed to exalt nationalist feeling, to elevate the idea of technological progress or colonial conquest, or to commemorate the wars of modern times. The station became a great picture-book of instructive images.

1
The central entrance portal at Helsinki Station, flanked by four pink granite giants bearing globes that light up the night. Architect, Eliel Saarinen; 1904–14. (Photo Roos, Finnish Architecture Museum)
2
Sculpture over the main portico of the central station in Zürich. (Photo Giegel, Swiss National Tourist Office)
3
Sculpture on the façade of Amstel Station, Amsterdam, 1939; architect, M. J. G. Shelling. (Photo AMA)
4
Detail of column capital at Metz Station, France, 1905–08; architect, Kröger. Friendship between peoples. (Photo Planchet, CCI)
5
Sculpture over pediment of the right-hand wing of the Gare de l'Est, Paris (built after the First World War), a pious memorial of Verdun; it was from this station that soldiers left for the Front. (Photo J. Javaux)

1

2

3

4

5

Temple of technology

Many 19th-century authors described railway stations as 'Cathedrals of the Modern Age'. This may sound like high praise for the architects. But architects, trapped in academic conventions, mostly did no more than copy, adapt or combine examples from past centuries. It was not they who were the innovators in the design of stations, evoking comparison with the cathedral builders. On the contrary, their role tended, with rare exceptions, to be the camouflaging of the station's functional reality, so as to hide from the public the industrial modernity of the railway. To quote Georges Tubeuf, 'Stations are designed by engineers. The architect only comes along later to decorate them.'

In the 19th century, a new sort of man, the engineer, with a totally different ethos and skill, appeared in response to the new needs of industrial society. The building of the railways offered engineers their greatest challenge. Above major station platforms, to protect travellers from bad weather and, crucially, to unify the space, were erected gigantic roofs whose physical – and spiritual – dimensions were the true source of the comparison with great churches.

These structures were of necessity lofty, to permit the dispersal of locomotive steam, but that they were built with such verve demonstrates a powerful act of faith in the future offered by technology and industry. They are still redolent with grandeur and a serene beauty resulting from the novelty of the design problems that were solved by their construction, and from the imaginative exploitation of the structural possibilities of wood, iron, glass and steel. So the engineer's station became the frontier of technical advance, and the symbolic embodiment of the promise of modernity.

Thus, in the 19th century, the railway station appeared to be the setting for a confrontation between the old and the new. While architects produced station buildings redolent of the past, the engineers gave vigorous expression to their faith in the future by erecting great sheds in apparent defiance of the past. However, since 1900 architects have gradually renounced historical pastiche to embrace functionalism, so that this opposition between architect and engineer no longer seems at issue.

The great stations which aroused the most controversy in this century, and which pushed railway architecture most firmly in the direction of functionalism, did not include the engineer's grand train sheds in their design – stations such as Helsinki, 1910; Cincinnati, 1929; Florence, 1933; Rome, 1947; and, more recently, Paris-Montparnasse. In these modern stations and many others, shelter for travellers has been provided by modest constructions only above the platforms. No overall roof has been built to give that sense of interior which marks King's Cross, St Pancras, or Paddington in London. It is true that with the disappearance of steam locomotives, grand train sheds are no longer necessary. But their demise more profoundly shows a shift of spirit. No more do stations so exuberantly celebrate the triumph of technology.

Opposite: the great hall giving access to the platforms in Pennsylvania Station, New York, 1906. (Photo Museum of the City of New York)

1
Wooden roof over the Santa Maria Antonia Station, Florence. (Photo FS)
2
Wooden roof over the first station at Temple Meads in Bristol, 1840. (Photo NRM)
3
Wooden roof of Central Station, Copenhagen, 1912. (Photo DSB)
4
Metal roof at Derby Station, England, 1841. (Photo NRM)
5
Metal roof of the first Grand Central Station, New York, 1869–71. (Photo Museum of the City of New York)
6
Metal roof of Barcelona-Termino Station, Spain, 1924: one of the last large metal station roofs built in Europe. (Photo RENFE)

4

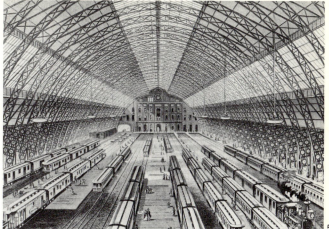

5

6

1
Metal roof of the station at Lille, France, 1889. (Photo SNCF)
2, 3
Interior and exterior views of the railway workshops at Baton Rouge, Louisiana, 1958. Engineer, Buckminster Fuller. The first geodesic dome constructed entirely in steel: diameter 117 metres. (Photos USIS)
4
Competition design for the concourse of Naples Station in Italy, 1954. Architects and engineers, E. Castiglioni, G. Bongiovanni and E. Sianesi. (Photo FS)
5
Locomotive Round House at Lyons, France, c. 1947. Engineer, Lafaille. (Photo SNCF)
6
Competition design for the main station in Naples, 1954, by Pier Luigi Nervi, in collaboration with Vaccaro and Campanella. (Photo FS)
7
Booking hall of the station at Karlsruhe, Germany, 1908–13. Architect and engineer, August Stürzenacker. One of the first uses of reinforced concrete in transport architecture. (Photo CCI)
8
Control cabin at Utrecht Station, Holland, 1938; architect, S. van Ravenstyn. (Photo AMA)
9
Communications tower, built by the Canadian National Railways in the yard of Toronto Station, 1973–76: height 553 metres. (Photo CN)

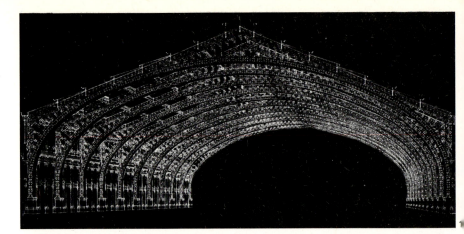

1
Competition design for the main station in
Bucharest, Romania, 1894; by Alexandre
Marcel. (Photo CCI)
2
Metal roof for the shed of the first Gare du
Midi at Brussels. Such was the airy grace,
indeed the majesty of the space created here
by the engineer that it was to be chosen,
before being brought into service, as the site
for the staging of royal entertainments.
(Photo ACL)
3, 4, 5
Central Station, Milan, 1913–30. This great
terminus is a late example of the classic
dichotomy in station architecture. Here a
megalomaniac assemblage of hybrid stylistic
elements, vaguely inspired by Assyro-
Babylonian sources, clashes with the roof
over the platforms, whose functional clarity
is the only element that evokes the modernity
of the place. (Photo FS)
6
Lytham Station, England, 1846. The first
stations illustrated a hybrid conjunction of
two opposing languages, two divergent
ethics. The sober wooden roof of the shed
over the platforms is concealed behind neo-
classical architecture.

3

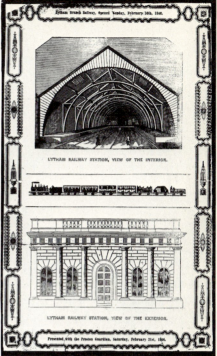

6

4

5

A case history in modern architecture

The birth of the railway gave rise to a new type of building which had to meet completely new requirements: not only to bring together in one single structure all sorts of different activities, but also to resolve the problem of moving travellers and goods to and from the new mode of transport.

As a result, the station experienced a long period of gestation before it acquired an architectural identity of its own. The rudimentary 'embarkation points' gradually gave way to pieces of architecture that attempted to produce some sort of harmony between the necessary interior spaces and the external form of the building. For example, the central feature of the entrance or booking hall was expressed in the façade by a colonnade or a line of arches.

This kind of rational approach failed to provide a rich enough language to satisfy the publicity needs of the numerous competing private railway companies. They each wanted a style to endow with their own image, to impress the public eye. So, from the 1850s onwards, there was a great flowering of fanciful architecture – Renaissance, Gothic and Romanesque – in station forebuildings which denied the technical reality of the railway and the presence of the train shed. This outraged the pragmatists, but their most striking suggestion – a station with a façade reflecting the structure of the roof behind it – was achieved only in a few rare examples.

Curiously, triumphant industry did not display with pride such features of its achievement as the train-shed roof or the locomotive itself, but hid them behind eclectic façades, the opulence of which was meant to reflect the wealth of the town and not to suggest any disturbing novelty. The new bourgeoisie could not yet admit that an industrial aesthetic, which it accepted and absorbed in its place of work, could be applied to buildings where something more spectacular was expected.

The separation between the true function of the building and its architectural treatment began to fade in the first decade of the 20th century, in a process disturbed only by a few outbursts inspired by ultra-nationalist governments. From now on all architectural styles with the least hint of romanticism were abandoned. Art Nouveau was to appear only in a few decorative touches (most charmingly in the Paris Métro). None of the expressionist or futurist designs for stations put forward just before the First World War was ever to see the light of day.

In the 1920s and 1930s there was disagreement between the station builders, resolutely modernist, and municipal authorities, anxious that their local identity should be preserved in traditional styles. At the same time, grandiose and triumphal architecture disappeared completely from European railway stations in the interests of a more rational use of space.

The new international style, the concrete revolution, postwar austerity and, worst of all, the steady elimination of the steam locomotive dealt a fatal blow to the image of the railway station which, in fact, has lost most of its identity since the Second World War.

Opposite: Gare des Bénédictins, Limoges, 1925–29; architect, Gonthier. The outcome of a series of projects from 1908 onwards, this station's siting and its backward-looking architecture excited passionate local controversy. (Photo Planchet, CCI)

1
Gare de l'Allée Verte, Brussels, 1835 (now demolished). The first station to be built in continental Europe, it had a long life in spite of the provisional appearance of its wooden architecture. (Photo SNCB)
2
Plan for a railway embarkation point at Meaux, c. 1848; architect, Arnoux. The use of the term 'embarkation point' for the earliest railway stations in France reveals the model to which builders turned: canals and river navigations. (Photo CCI)
3
Willemspoort Station, Amsterdam, 1842–43; architects, F. W. Conrad and C. Outshoorn (closed in 1878). In semicircular form, this type of station with its references to neo-classical temples was commonly used following the first French 'embarkation points' at St-Germain, Sceaux, etc. (Photo SMU)
4
St Pancras Station and hotel in London, 1868–76; architect, Sir Gilbert Scott. This station was much admired in its day for the harmonious union between its forebuilding, incorporating a hotel, and the train sheds. In order to break the monotony, the architect used a Gothic Revival style, accentuating the picturesque with its towers and pinnacles. St Pancras remains an example of a station that is both ingenious and spectacular. (Photo NRM)
5
Gare du Nord, Paris, 1861–64; architect, Jacob Hittorf. This station replaced an earlier embarkation point built in 1846 by the engineer Léonce Reynaud. The façade of the new station had at the same time to give an indication of its purpose and to convey a monumental character appropriate to the importance of a major Paris station. The architect made a large number of proposals which ended in this compromise between rationalistic architecture and neo-classical styles. (Photo Roger-Viollet)

4

5

1
Station at Tours, 1895–98; architect, Victor
Laloux, a well-known professor of the Ecole
des Beaux-Arts who also built the Gare
d'Orsay in Paris. At the time, this structure
was considered by architectural theorists to
be the perfect expression of the station, as
the Gare de l'Est in Paris had been before.
(Photo Roger-Viollet)
2
Central Station, Helsinki, 1910–14; architect,
Eliel Saarinen. The result of a competition in
1904, this station inspired a national debate
between partisans of national-romantic and
academic styles. Intended to be the
embodiment of a nordic tradition, the station
is an amalgam of academic, romantic and
rationalistic features. However, the unity it
achieves makes it a masterpiece of railway
architecture of the period. (Photo Havas,
Finnish Architecture Museum)
3
Metz, City Station, 1905–08; architecture,
Kröger. Built during the German occupation,
this station excited much criticism in spite of
its good features because of its Romanesque
style, then common in Germany but foreign
to this part of France. (Photo A. Schoutz,
SNCF)
4
Central Station, Antwerp, 1899; architect,
Louis de la Censerie. Eclectic architecture
run wild. (Photo AAM)
5
Sketch for a station, 1914; architect, Erich
Mendelsohn. The German Expressionists
sought to symbolize the function of a
building in its form. Their few station
projects, deliberately disregarding the
technical constraints of railway requirements,
have a surprisingly romantic force in their
highly subjective forms. None of these
stations was ever built. (Photo CCI)

1

2

3

4

5

1, 2
The station at Deauville, France, 1930;
architect, Jean Philipot. The Norman style
imposed by the town reflects the movement
towards architectural regionalism in the
1920s and 1930s in France, a movement
also affected railway stations. Deauville
Station was illustrated in many architectural
manuals and was even copied in the French
colonies. (Photos SNCF and CCI)
3
Hall of Milan Station, 1920–30; architect,
Ulisse Stacchini. Italian stations,
distinguished up to this time by their well-lit
interiors, but not otherwise particularly
innovatory, were to cause an international
stir between the two World Wars. Under
Fascism, public buildings had to glorify the
reigning ideology through an imposing and
monumental aesthetic, on the borderline
between the magnificent and the grotesque,
as here at Milan. Almost at the same moment
the Santa Maria Novella Station in Florence
was making architectural history.
4
Santa Maria Novella Station in Florence,
1934–36; architects, G. Michelucci and
associates. A victory of modernism over
grandiloquence, the stark forms of this
station evoked strong reactions in its users;
comparisons were made to a piano, a
luggage trunk or even to a dam. (Photo FS)
5
Station at Lens, 1926; architect, Urbain
Cassan. Rejecting the architect's first plan as
nothing but a 'market hall', the city called for
'an articulated architectural structure', to be
compatible with the whole of the square on
which the station was to stand. This demand
was never satisfied because the architect,
faced with the problem of building over mine
workings, sought to make the structure as
light as possible. (Photo CCI)
6
Heating centre and control cabin of the
Santa Maria Novella Station, Florence,
1928–32; architect, Angiolo Mazzoni.
(Photo Anderson, FS)
7
Project for Central Station, Rome, 1939;
architect, Angiolo Mazzoni. An example of
the triumphal architecture typical of the
Fascist period. (Photo AM)

5

6

7

1
Stazione Termini, Rome, 1948–51;
architects, L. Calini, A. Pintonello. Towards
the end of the Fascist period only the
arcaded side buildings (not visible here) had
been completed by the architect Mazzoni. In
1947 a competition was launched for an
edifice 'to reflect the new style of life in
Rome and the change in political ideology'.
The entrance hall, with its S-profile roof in
reinforced concrete and glass, constituted
the most successful element of the station
and signalled the rise of a new school of
Italian architects, of whom Pier Luigi Nervi
was best known, who worked closely with
engineers. (Photo FS)

**The architectural history of railway
stations is marked by struggles
between modernist efforts and
periodic revivals of the traditional
language. This conflict is illustrated
even within the work of an individual
architect: for instance, Van Ravesteyn
in the Netherlands (inspired by the
forms of Baroque and neo-classical
architecture) and, earlier, Mazzoni in
Italy.**

2
Central Station, Rotterdam, 1957; architect,
Sybold Van Ravestyn. (Photo Spies)
3
Station of Vlissingen, Netherlands, 1950;
architect, Sybold Van Ravestyn.
(Photo AMA)
4, 5
Principal façade and waiting-room at
Volgograd (formerly Stalingrad) Station,
USSR, 1953; architects, A. Khourovskin
and S. Briskin. (Photo Soviet Railways)
6
Project for a building in front of Milan
Central Station, 1952; architects, G.
Minoletti and E. Gentili. For a long time the
Milanese were embarrassed by the station,
not so much because of its heavy and
pretentious façade as for the Fascist
memories it perpetuated. (Photo FS)

1

2

3

4

5

6

Hub of the city

Stations played a key role in urban development. The mouth of a channel for large flows of goods and people, the station was, from the first, a point around which the modern city grew.

Stations have moved the centre of gravity of many cities from their historic cores and have caused the appearance of new patterns of city building and new concentrations of residential building, all of which has greatly affected the life of city-dwellers.

As a node of the railway network, the station accelerates and directs the growth of cities, and even of giant metropolitan areas. In the USA, Chicago owes its meteoric rise and rapid physical growth to its position as the meeting-place of twenty-seven different railway lines. In the new economic logic of transport and communications of the 19th century many a sleepy town and village, endowed by the chances of railway geography with a marshalling yard, became a regional economic centre. Conversely, many an old and prosperous town, deprived of railway contact, saw its prosperity suddenly threatened. In fact, during the last century the nature and location of new urban settlements were to a great extent determined unwittingly by railway planners. When the engineers laid out their lines leading to cities, they were influenced by technical considerations, the need for economy and competitive factors — not by concern for the ordered growth of the city.

A notable result was the chaotic mushrooming of unplanned, ill-serviced suburban developments dependent on urban centres at ever-increasing distances. So we see the inherent inability of industrial society to coordinate and harmonize two essentially complementary developments, resulting in the daily Gadarene migration between suburbs and city centre.

However, there were exceptions, and among this haphazard growth one must note some experiments in town-planning carried out on the edge of great cities. In the mid-19th century, there was Vesinet to the west of Paris, Bedford Park to the west of London, and Riverside near Chicago. Later, the modern ideal of 'town-in-country' took the form of garden cities in the development of which the great protagonist was Ebenezer Howard. His first garden city at Letchworth in Hertfordshire, like his second, Welwyn Garden City, depended entirely on the railway. But it was meant to be self-supporting, with its own industry, and not a dependency of Greater London.

In 1912, a new grand plan for the development of an existing town, arranged on a railway network, was conceived for Helsinki, where Eliel Saarinen proposed a group of small satellite communities, each surrounded by greenery and dependent on its station. The plan was not carried out to any great extent, but its principles were taken up again in a modest way in the 1920s and 1930s in many places, and notably in 'Metroland' to the northwest of London. The British new towns of the postwar period, in spite of the enormous increase in road transport, were all sited on major railway routes.

Opposite: caricature by A. Robida, 1896. A view of Paris showing 'elements of beauty and dynamism' which the railway 'can bring to the landscape of large cities, the admirable transformations which it can effect and finally how it makes ingenious and picturesque use of monuments which until now have served no purpose' . . . (CCI)

At the end of the 19th century stations still provided a legitimate opportunity for town-planners to make an expression of grand monumentality and to create a powerful centralizing image in the city.

1
Project for an unidentified central station for a German city, 1911. (Photo Bouchart, CCI)
3
Project for a central station for Paris located between the Place de la Republique and the St-Martin canal, 1903. (Photo Planchet, CCI)

Between the two World Wars, and with the emergence of competitive forms of transport, the station lost its triumphal image, but for a long time still it remained a functional hub of the city.

2
Project for a new town, 1922; architect, Le Corbusier. Though the station is still the geometric centre of this agglomeration of 300,000 people, it is already relegated to an underground level; it disappears from view and has no architectural expression in the silhouette of the town. It is significant that the underground complex is covered at upper levels by the path of a motorway — which forms the visible axis of the city — and an aerodrome. It is this kind of transport interchange that now becomes the key to the organization of the city.

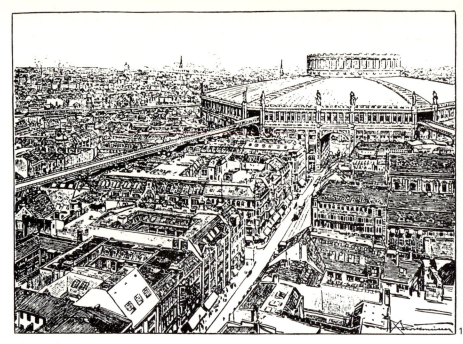

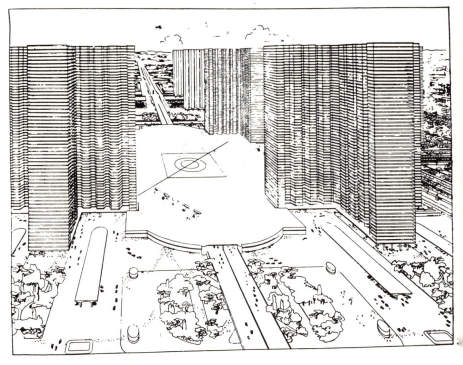

3

As much from the town-planning as from the psychological point of view, the station became a new entrance to the city, taking the place of the gates in old city walls. This threshold aspect of railway stations was expressed in multiple variations on the themes of portico and triumphal arch.

1, 2
Entrance to the Central Station in Zürich, derived from the triumphal arch. (Photos ONST and the Archives of the City of Zürich)
3
Triumphal-arch entrance gate to Tesnov Station, Prague. (Photo V. Slapeta)
4
Project for an entrance gate to the Gare de l'Etat, Saumur, France. (Photo CCI)

2

3

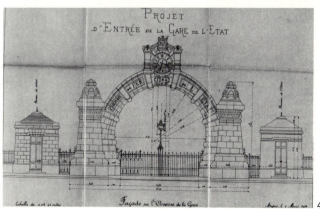

4

The neighbourhood of the station is
perhaps the only part of the city which
shows its bad side and its good side
straight away. Seen from the platforms
or the trains, it often displays a long
procession of back yards, nondescript
pieces of ground and decrepit
industrial installations – the sordid
behind-the-scenes of the modern city.
The area of the station frontage tends,
by contrast, to have architectural
trappings intended to impress the
arriving visitor with the appearance of
an elaborately ordered urban setting.

1
Competition scheme for the arrangement of
Lucerne Station, Switzerland; architects,
Werner Kreis and Ulrich Schaad. (Illustration
Kreis and Schaad)
2
Competition entry for Leipzig Station, 1907;
architect, Peter Birkenholz. (Photo AAM)
3
The Place Marie-Henriette in Ghent,
Belgium, in front of the St-Pierre Station,
1908–12; architect, Cloquet. (Photo IRPA)
4
Competition design for Karlsruhe Station,
1904; architect, Hermann Billing. (Photo
AAM)
5
Piazza in front of Central Station, Milan.
(Photo City of Milan)

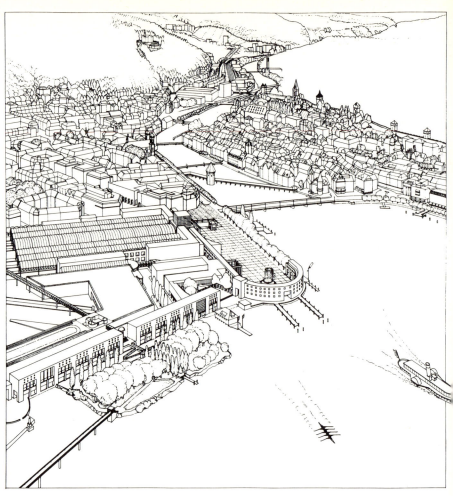

3

4

5

The role of the station as the hub of urban development looks set to continue. There are numerous contemporary examples in many different countries of vast undertakings in town-planning and urban renewal designed around stations. Their nature and scale are such as to affect profoundly the centres of gravity and the circulation patterns of the cities where they are located.

1
The new *'centre directionnel'* at Lyons, built next to the Gare de la Part-Dieu. (Photo Studios Villeurbanais)
2
Model of the new business and commercial centre being built around the Central Station in Utrecht. Note the brutal opposition of scale and architectural treatment between the traditional urban centre (in the background) and this new complex. (Photo Municipal Archives of Utrecht)

In large cities the enormous areas of railway land around stations constitute one of the last reserves for building vast urban projects without involving the destruction of historic buildings or of green spaces.

3, 4, 5
Aerial view of the land occupied by the station in the middle of the urban fabric of Toronto, with a view from the same angle of a model of the proposed new city centre to be built on the site. Conceived as 'the most grandiose project of urban renewal ever undertaken in the heart of a North American city', this programme underlines the stakes involved in station developments today. The operation is estimated to cost in all one billion dollars, and is to be completed in fifteen years. (Photos Canadian National)

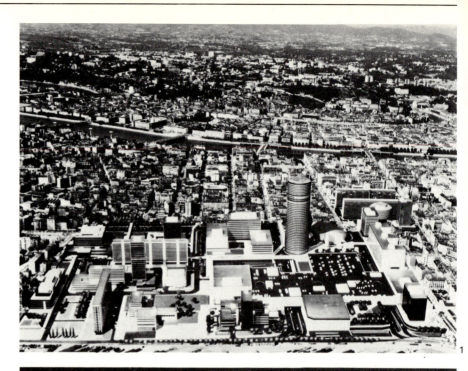
1

2

3

4

5

The insertion of a central station and a new rail transit system (between the Gare du Nord and Gare du Midi) into the heart of the historic city of Brussels will probably have the most drastic town-planning and social consequences of any such project. To assure the rapid movement of trains between the north and south of the country (and between other European countries), extensive demolitions have been and are being carried out across the whole of the old city. The government has acquired the enormous land holdings involved, and is implanting on this axis – straight through the medieval city – continuous alignments of office buildings that are cutting the old city in two and that seem certain to engender, by osmosis, an increasingly brutal 'urban renewal'.

This operation, guided by purely technocratic criteria, is precipitating a marked depopulation of the central areas and could lead to the gradual devitalization of the inner city. On the other hand, Brussels will have a rail transport system of unrivalled efficiency.

1, 2, 3
Three views of the violent gash across the medieval centre of Brussels, caused by the construction of the rail link between the Gare du Nord and the Gare du Midi. (Photos Bouchart, CCI)
4
Aerial view of the centre of Brussels with the locations of successive stations: (1) first station in the north of the city, 1835, demolished; (2) first station in the south, demolished; (3) second Gare du Nord, demolished; (4) second Gare du Midi, demolished. Between the two last, there was created in the 19th century a single great urban boulevard across the whole city, on the model of Haussmann's avenues in Paris. Many plans were considered for creating a railway link between the north and south under this boulevard, with a central station in the middle (5); these plans were never carried out. (6) The third, existing, Gare du Nord; (7) the third existing, Gare du Midi. Between these two stations was built the 'north-south rail link' punctuated by two secondary stations (9, 10) and a central station (8), whose surroundings (11) are still, twenty-five years later, brutally scarred from the operation of dismantling the city centre. A vast open car park is now situated in the heart of the city. In the 1960s an ambitious urban renewal project between the Gare du Nord (6) and the goods station (12) was started. Its failure contributed still further to the image of Brussels as a city devastated by the combined activities of technocrat-planners and property speculators. (Photo IGM)

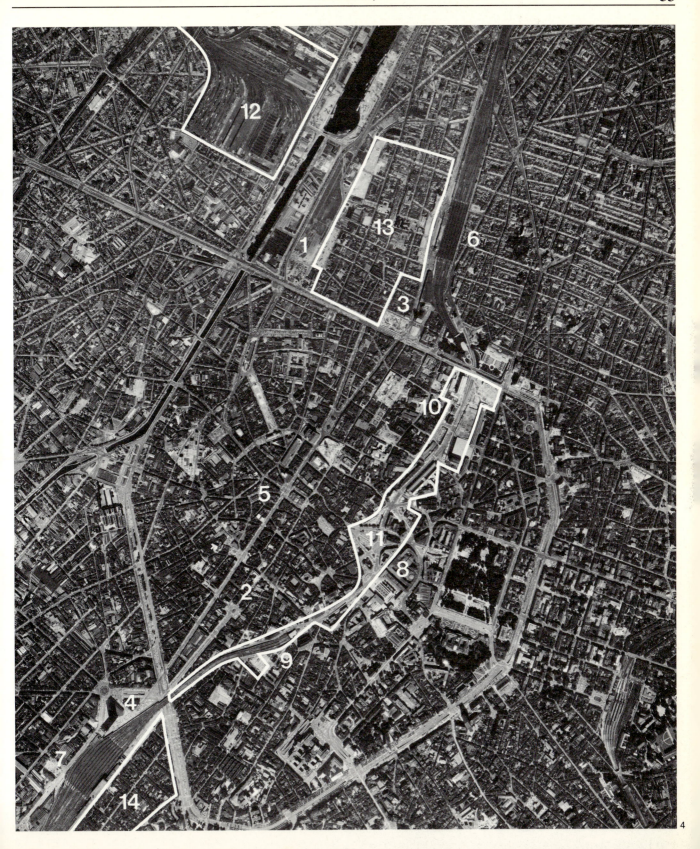

1
Between 1920 and 1930 there was an attempt, in the London area, to control suburban growth by planning housing projects around the stations serving the new railway lines created for this purpose. Here, deep in the country, is the site reserved for Edgware Station in 1923. (Photo Museum of London)

2
The coming of the railways led to ideas of urban decentralization which were promoted by Ebenezer Howard as early as 1898. He proposed an alternative between the traditional city and the country: the idea of the garden-city which would offer an ideal compromise for the growth of modern society. While these cities would be economically autonomous, they would be dependent on rail communications for assuring their organic links with the country at large. (Photo CCI)

3
Plan for the creation of a 'new town' to allow for expansion of the city and port of Antwerp, 1856. The station here is conceived as an element of a new phase of urban growth, as the kernel of a new economic structure of the metropolis. (Photo Royal Library, Brussels)

4
In an attempt to overcome the chaotic growth of suburbs, which had been developing since the 19th century along main railway lines, town-planners around 1920 had the idea of segregating industrial and residential development along different lines. The theory was, however, very rarely applied in practice. (Photo CCI)

5
The growth of the London conurbation from the beginning of the railway era to the present. (Photo CCI)

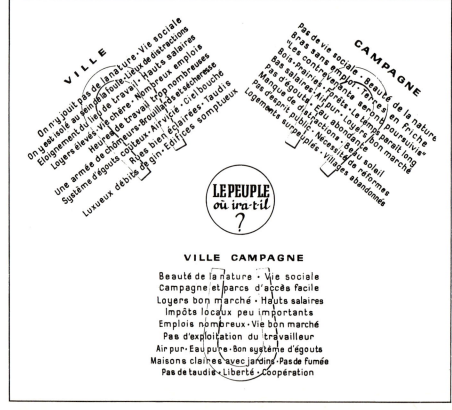

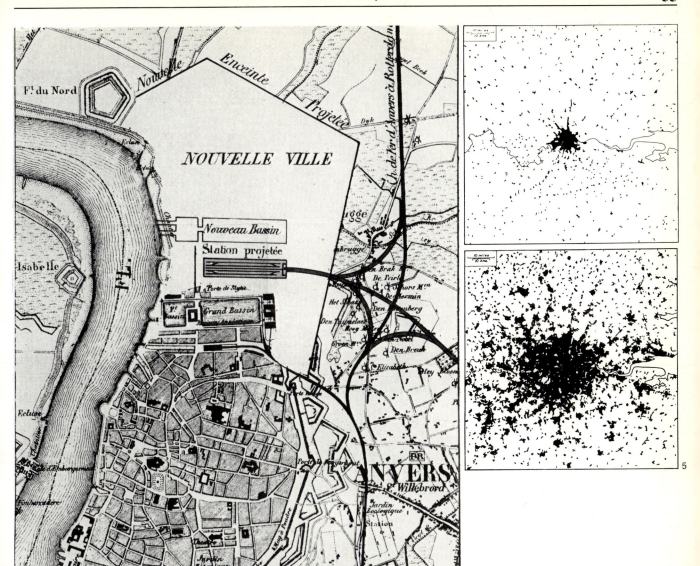

3

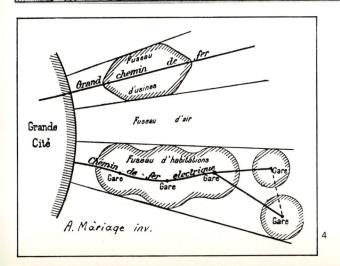

4

5

1
Plan for a model town to be built by railway companies following the taming of the American West. The station is here the geometric centre of a system of streets to be repeated in identical form in each settlement along the line. (Photo CCI)

2
The construction in the USA of trans-continental railway lines involved the lodging and management of an enormous labour force which had to be moved as the work advanced daily into deserted regions. The men were housed in trains of three-storey dormitory wagons which seem to anticipate some of the visionary ideas of 20th-century planners for linear cities or for cities on wheels. (Photo Association of American Railroads)

3
Plan for a complete new town around a station of the Swedish railways, 1859; architect, Edelsvärd. (Photo CCI)

4
Aerial view of Herington, Kansas, 1887. The railway station appears as the nucleus of the new town. (Photo Library of Congress, Washington, D.C.)

5
The euphoria aroused by the arrival of the railway in formerly isolated regions of the American West gave birth to numerous settlements which had no other justification than the speculative visions of their promoters. A number of them were abandoned soon after they were founded. Here is a railway ghost town in Kansas, 1871. (Photo CCI)

6
In the 19th century, in order to promote the creation of railway lines across the American continent, the government made tax-free grants to railway companies of vast territories for exploitation on either side of the planned routes. The total area of these concessions was estimated to be 131 million acres. The upper map shows in true proportion the size of these land grants, while the map below corresponds to exaggerated information circulated at the time. (Photo CCI)

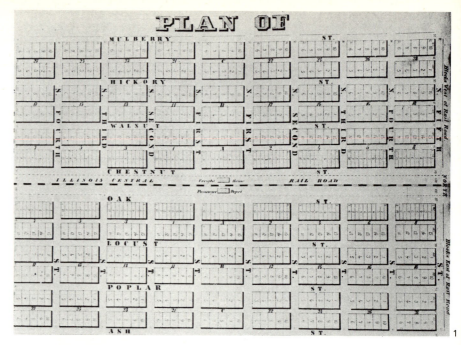

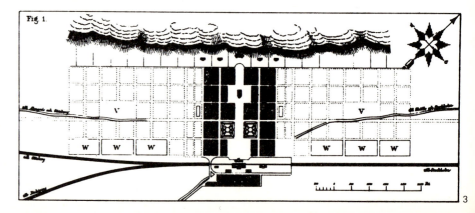

Microcosm of industrial society

Product and reflection of class-ridden society, the railway station was one of the first places in modern times where people of all classes rubbed shoulders. These cathedrals of the Industrial Revolution resounded with the glad tidings of the religion of progress brought by the railway, and they proclaimed the ascendancy of the new middle class with their ostentation and pride. A sort of railway aristocracy took their place in the 1920s, a time when great trains – and luxury travel in general – became enshrined in myth.

Through the monumental façades of great town stations, via little stations in the country, whole populations moved in the 19th century from the agricultural regions towards the factories and the offices of industrial towns. For these people, who became labourers or white-collar employees, the station no longer meant anything more than an obligatory daily event, passed through twice a day in the journey from home to work and back again.

The general introduction of paid holidays brought the same crowds through the stations creating their characteristic image as settings for departing and returning holidaymakers.

The jet-set has dethroned the railway gentry, and the romantic adventure of the great international trains in the vintage years has given way to the dreary regimentation of the airport. Now it is the migrant workers from Yugoslavia or Turkey who ride the (second-class) seats of the successor to the Orient Express (renamed Direct-Orient). Unlike the travellers of the past, they probably see their train as no more than commuter transport writ large. All the same, in the eyes of the exiles who alight from these trains can be seen the same uprooted look that marked the Italian or Alsatian emigrés who, in the 19th century, haunted the Gare St-Lazare in Paris, waiting for the boat-train to Le Havre and the ship to America. It is extraordinary to see the station of a typical European industrial city transformed on weekends into the meeting places of migrant workers. For them the station becomes the nearest link with their home countries in a society which is prepared to use but not to accept them.

But railway stations, being places of passage, are also the asylum for others whose travelling is finished: people on the fringe of society, tramps, dropouts, the subject of another mythology of the railway station often represented in fiction and the cinema. Around them the station benignly goes about its business.

Opposite: weekend departure from Glasgow Central Station. A picture of the 1930s; but the event has become ritualized – the tumult of crowds fleeing the cities where they work. (Photo City of Glasgow District Council)

Because stations are places where large crowds pass through, they often reflect the movement of ideas and the eruptions of social stress. They have also served as ideal places for propaganda and political argument, because of the captive audience which they offer.

1
Graffiti in a suburban station at Sarcelles, France, 1978. The slogan reads: 'Women, don't imitate, invent!' (Photo Planchet, CCI)
2
'The Great Socialist Revolution', USSR, 1917. Political meeting in a Soviet station. (Photo Snark)
3
Crowd gathered in front of the Gare de Lyon, Paris, May 1968. (Photo B. Barbey/Magnum)

2

3

1
Munich Station in Germany. Munich is the
distribution centre for Turkish immigrants
who have come to work in Germany, 'the
new slaves of Europe'. The process of sending
them to their destinations is organized in an
old air-raid shelter within the station. (Photo
Gilles Peress/Magnum)
2
A station in Paris, 1978. Waiting and
boredom. (Photo C. Louis)
3
Cologne Station, West Germany. Turkish
workers on their way home for the holidays.
(Photo L. Freed/Magnum)

2

3

1
'The soldier's farewell', Milan Station, 1960s.
(Photo B. Barbey/Magnum)
2
Entrance to the cemetery station at Waterloo,
London, about 1850. This station, specially
constructed by the London Necropolis
Company, was connected to the vast
cemetery at Brookwood, near Woking,
twenty-five miles from London, by a special
service of funeral trains. (Photo NRM)

2

A place of order and discipline

The cult of punctuality was born with the railways. Before rapid travel was made possible by the railways, regions worked to their local time. In England, for example, midday by solar time comes a quarter of an hour later in Plymouth than in London. But railway timetables based on local times were confusing: why should it seem to take longer to travel from Plymouth to London than from London to Plymouth? Railway time was the precursor of Greenwich Mean Time, a national standard required to produce a rational timetable as soon as the railway network developed from a series of local lines into a national system. The railway introduced the idea that time is money. In Tokyo, suburban services are timed not in minutes but in seconds.

The organization of the railway station was originally on military lines. There were Orders of the Day, a hierarchy of authority, fines, suspensions, and even uniforms adjusted to status like those of an army. Mutiny was also not unknown. The station might be the setting for spectacular resistance by the railway workers, either against their own authorities or against the military forces of an occupying power.

Being concerned with maintaining the movement of masses, the station was one of the first places in which crowd control was practised. It was also perhaps the only public place where the social divisions of class were built into the structure, with distinct facilities for each category of traveller. Nowadays this is less marked, but there is still an element of segregation. There is also a strong suggestion of policing. Special legislation enshrines the bye-laws, the things which are forbidden, the misdemeanours, and the fines. There is also a railway police force.

Opposite: a Dutch socialist propaganda poster, 1903. A powerful worker puts a stop to the functioning of the machine with his ultimate weapon, the strike. (Photo AMA)

'Railway time' coexisted for many
years with local, solar time, as it
appeared on public clocks and church
towers. In the second half of the 19th
century the two times were brought
into coordination, mainly through the
railways, since travellers had to adapt
to scheduled arrivals and departures.

1
'The customer's point of view.' Circular
issued by the French PLM Railway, 28 July
1933. The cult of precision and the idea of
public service join in the railwayman's ideal:
'on time'.
2
'Railway Time', c. 1850. Station clock
distinguishing railway time from solar time.
(Photo NRM)
3
Ritual image of the religion of punctuality:
an employee of the Great Central Railway
sets the clock indicators showing train
departure times. (Photo NRM)
4
Pennsylvania Station, New York. The clock
dominates. (Photo USIS)

**The station tower with a clock on each
face is an architectural archetype,
functional and symbolic at the same
time, which has known many
variations.**

5
Tower of Helsinki Station, 1910–14;
architect Eliel Saarinen. (Photo Finnish
Architecture Museum)

*'There is even a railway time followed by the
pendulums as if the sun itself had given up.'*
Charles Dickens, 1848 (about the Charing
Cross Station area in London).

2

3

4

5

1
Railway workers were notoriously subject to petty discipline. In this design for a poster (France, 1944) a train driver shows off his good conduct record. The notion of 'public service', strong in some countries, makes a virtue of obedience. The necessary 'high morality' of the railway worker complements the obligations of discipline.
2
Staff of South Station, Stockholm, c. 1900. Even the pose for posterity respects the levels of the pyramidal hierarchy at the station. (Photo Swedish Railway Museum)
3
The great American strike of railwaymen, 1877. The station is often the scene where railway workers chose to express their opposition to exploitation. Shown here is Pittsburgh Station and hotel set on fire by strikers. (Photo Library of Congress)
4
Seat of the 'Dopolavoro Ferroviario' in Rome: entrance to the theatre, c. 1930. In Mussolini's Italy, this institution took charge of railwaymen's leisure time, insidiously extending the railway organization's control of them. (Photo FS)
5
Staff uniforms of the Saint Germain and Versailles Railway, c. 1840. Railway uniforms were modelled on those of the military; general rules about dress were part of the strict code governing the behaviour of railway employees. In certain countries, station masters for a time even had the right to carry a sabre. (Photo VDR)

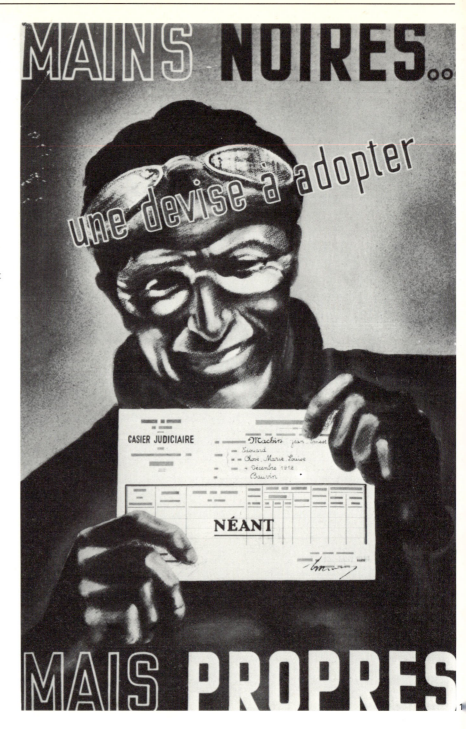

2

5

3

DOPOLAVORO · FERROVIARIO · DI ROMA

4

1
The 'crows's nest' of the Gare St-Lazare in Paris. The essential observation point for the control of crowds in a station which sees 115 million travellers each year. (Photo Bouchart, CCI)
2
Caricature, 1884: the treatment reserved for travellers according to their social origins. (Photo RENFE)
3
Cover of a novel prohibited from distribution to bookshops on the Orléans line as 'contrary to good morals'. (Photo Bouchart, CCI)
4
The other side of the curtain, 1978. . . . these marvellous places from where one sets off for a distant destination are also places of heartbreak,' Marcel Proust, *Remembrance of Things Past*. (Photo Maïofiss)
5
Bengal Express, 1910: fifteen minutes' stop, the barber on the platform shaves the passengers. (Photo Almasy)
6
Myth and reality in the Third World: a train taken by assault, Lagos-Kano line in Nigeria. (Photo Almasy)

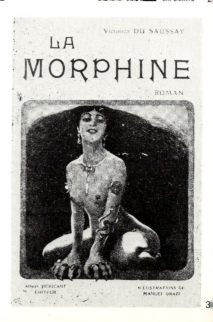

4

5

6

The station: strategic area

Many early stations on the Continent were placed outside city walls for reasons of traditional military strategy. The demands of defence remained, if only for a short while, more important than the convenience of movement.

Meanwhile, army leaders soon came to appreciate the military potential of the new means of transport. The American Civil War (1861–65) showed this for the first time. In France, after the confusion of mobilization in 1870, the Ministry of War put out a series of orders relating to the organization of railways in wartime. In 1913, a detailed plan was worked out for railway stations to serve specific strategic purposes. There were mobilization stations, control stations, depot stations, stations designated for the evacuation of the wounded, stations for men on leave. The control stations, whose possibilities and limitations were to be fully discovered during the First World War, were in fact nerve centres for the distribution of troops along the fronts. Even after the war this organization was continued on a reduced scale under the control of an officer nominated by the private railway companies.

During the Second World War, the enlarged theatre of operations and the increase in armaments enhanced the crucial strategic importance of railways. At the same time, air blitz had become a devastating new weapon. Railways were particularly vulnerable; stations, junctions and marshalling yards became key targets. By the time of the Liberation, for example, two-thirds of the railway stations in France had been destroyed.

It was perhaps above all in railway stations that the classic scenes of wartime drama were played over and over again. It was on the platforms, and at the windows of trains departing for the front, that one saw the smile of confidence in victory. Often, chalked on the carriages, were the supposed destinations of a victorious army: 'Next stop Paris' (in Germany) or 'Next stop Berlin' (in France). But behind this brave front lay the lingering glances and poignant farewells, the separations, the patient waiting and hoping of the womenfolk left behind.

Later these platforms might see the same men returning, some wounded, some dead. They might also see the passage of prisoners of war, of refugees fleeing before an advancing army. Stations received victors and vanquished alike.

Even in peacetime, these memories remain and it is impossible, if one has the time to linger in a great railway station, not to feel the weight of human emotion which has been expressed there. Some stations remind us in their decoration of the role that they have played. One can find it in Waterloo Station, in the triumphant ornamentation of the station in Milan, and in the Gare de l'Est in Paris, which was awarded the Legion of Honour, as if the station itself were a living hero.

Opposite: German military advisers to the Japanese on the platform of a station in China, 1938. (Photo Capa, Magnum)

1
An opening made in the fortifications of Antwerp for the railway, 1875. The fear of a Dutch invasion delayed this penetration of one of Europe's most strongly fortified cities. (Photo AAM)
2
Station at Adinkerque, Belgium. Above, left: as a country station in 1914. Below: the military extensions of the station made by the railway division of the Belgian Army in the course of four years of more or less static warfare. (Photo Coll. G. Neve)
3
The battle of City Point, Virginia, July 1864. The American Civil War demonstrated for the first time the key military importance of railways. In a fluid situation, neither side wanted the other to capture crucial railway installations, so these were destroyed on retreat. Being made of wood, stations were easily dismantled.
4
Intensification of bombardment on the station of Vaires, Paris region, between 28 June and 18 July 1944. The severe and repeated aerial bombardments showed the importance attached to destroying strategic railway centres serving the theatre of operations. (Photo CCI)

'It follows from all the discussions about the placement of the station in the city of Luxembourg that the character of this city as a fortress occupied by a foreign power (which had no interest in the growth of its prosperity but merely in the strategic value of its fortifications) was a very real obstacle to the plans of its citizens, who wished to make their capital a centre of lines of communication.

'The Prussian authorities, on military grounds, pressed for a long detour for the railway lines, and the station was constructed of wood.' (1858)
(Cahiers Luxembourgeois, 1953)

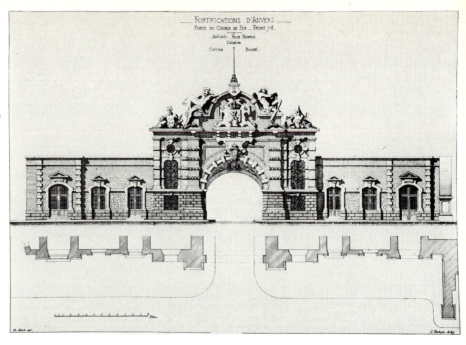

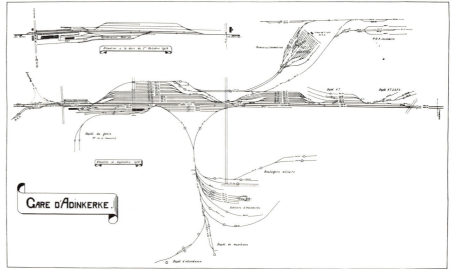

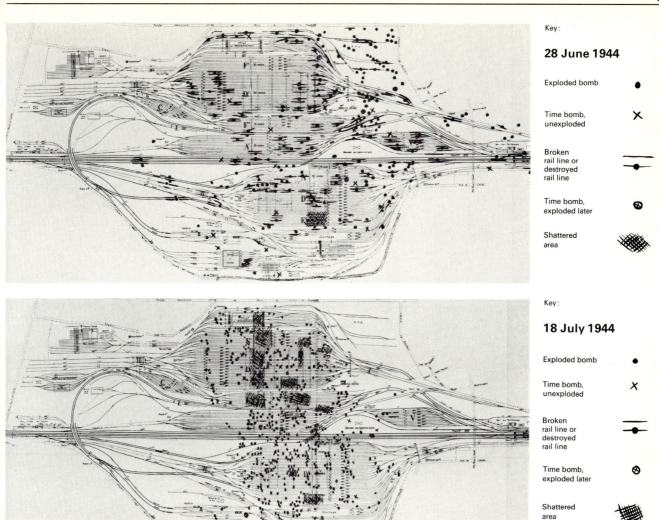

Key:

28 June 1944

Exploded bomb ●

Time bomb,
unexploded ✕

Broken
rail line or
destroyed
rail line

Time bomb,
exploded later ⊗

Shattered
area

Key:

18 July 1944

Exploded bomb ●

Time bomb,
unexploded ✗

Broken
rail line or
destroyed
rail line

Time bomb,
exploded later ⊗

Shattered
area

4

1
Gare d'Orléans during the siege of Paris. To
assure the defence of the city (1870), the
depots were transformed into workshops
where balloons were manufactured and
wheat was milled. (Musée Carnavalet, photo
CCI)
2
Workshop in a Paris station, 14 April 1917.
Women, mobilized in all areas of production,
cleaned and maintained equipment and
manufactured shells while the men were at
the Front. (Photo SNCF)
3
Americans passing through a station in the
suburbs of Paris during the First World War.
(Photo USIS)
4, 5
First World War, German and French
mobilization. The two opposed sides echo
each other in perfect symmetry. 'They all
sang going off to war.' (Photos Historical
Museum, Frankfurt, and René Dazy)
6
Hitler greeted by a crowd as he passes
through the station of Wilhelmshaven in
February 1936. (Photo CCI)
7
Behind the mask of heroic virility, the
anguish of parting and the prospect of a long
separation. (Photo CCI)

'There were more soldiers than we knew
what to do with,
There were two or three civilians, a father,
two mothers,
Who dried their fine eyes full of tears
While they said goodbye to their
grandchildren.
And there were soldiers pissing out of the
doors.
The night was calm above the train,
The locomotive was ready to depart,
Victory shone out from the eyes of the
conscripts.
Perhaps happiness is only in the Stations?'
Georges Pérec. (Translated from 'Whose is
the Little Bicycle with Chromium Handlebars
at the Back of the Yard?' (Denoël, 1966))

'The beloved
On the step of the carriage
Attaches to the passing hand
A thread wrapped around her heart.
When the train leaves, it empties
All her heart, the beloved dies.
Dead, she must go away
From the station, from the empty world.'
J. Cocteau, 'Farewell to the Marine Fusilier',
1918.

4

5

7

6

1, 2, 3
The celebration of the victory of 1918 is
superimposed on the memory of the great
English victory of Waterloo (1815) on the
façade of the station of that name in London.
(Photo Bouchart, CCI)
4
Returning prisoners-of-war, Vienna Station,
1945. A prodigal son comes home, while a
desperate mother seeks to glean news of her
missing son. (Photo Ernst Haas/Magnum)

4

Mardi nuageux

France-Soir

Allemagne 1 DM ● Angleterre 20 pence ● Belgique 12 F belges ● Espagne 35 pesetas ● Italie 350 lires ● Luxembourg 12 F luxemb. ● Pays-Bas 1.25 florin ● Suisse 0,90 F suisse ● Maroc 1,60 dirham ● Tunisie 120 mil. ● U.S.A. 70 cents ● Canada 75 cents

Paris, mardi 27 décembre 1977
100, rue Réaumur, 75002 — 508.28.00

1 F 40

BTD

bourse-courses toute dernière

Des autonomistes corses avaient dit : « Nous porterons la violence sur le continent »

La gare de Villepinte plastiquée à l'aube

Le F.L.N.C. revendique également la destruction de la villa du député Griotteray, près de Bastia

The station in politics

The idyllic vision of the railway as a link between peoples and the symbol of fraternity did not long disguise the essential nature of railway development, which was to gather an enormous amount of capital and make it profitable. The bankers realized this very early when they founded many of the first companies which were to become a powerful force in national economic life. They applied a philosophy of profitable territorial conquests, with important consequences for the areas served by the line.

In what were then emergent countries, such as the USA and Russia, the arrival of the railway and the building of a station caused towns to spring up. In Europe, the arrival of a railway station changed small towns into the effective centres of their regions, places where power and production were concentrated following the new logic of capitalism. In France, some secondary railways were built for immediately political reasons. 'Electoral Railways' strengthened the power of Napoleon III in relation to provincial landowners; while the stations in the Freycinet plan of 1879 were intended to carry the ideology of the Third Republic to every minor administrative centre.

In large cities, the locations, character and connections of railway stations were powerful factors in urban control. In Paris, Haussmann — carrying out his 'strategic embellishment' of the city — decided to provide the stations ranged round the centre with a series of large avenues radiating out from the inner core.

The impossibly grandiose scheme for Berlin demanded by Hitler in 1937 envisaged two main stations at either end of a vast avenue, the only purpose of which was to overpower the traveller and the visiting dignitary by its sheer scale and the grandeur of the urban perspective it revealed. It is impossible not to see, behind this ostentatious show of totalitarianism, the haunting pictures of Jews shepherded onto platforms before being sent to the 'terminuses' of the concentration camps.

High-level decisions which resulted in massive disruption of the urban scene have also aroused opposition and much political argument, making the station a focus of demands for 'urban democracy'. This occurred in Brussels, when the connection between the Gare du Nord and Gare du Midi was being planned, and in New York in response to the threat to Grand Central Station.

Apart from the political factors involved in their actual location, stations have become in themselves obvious centres for conflicting political propaganda. They were concentration points of militant effort on the journeys of the propaganda trains which helped to build Soviet Socialism. They were the scene of electoral speeches by American political candidates from their campaign trains.

The station at Villepinte, considered symbolic of French colonialism, was blown up by the Corsican Independence Movement in 1977. However gloriously the symbolism of a new industrial society was presented in the 19th century in monumental stations, opened with great ceremony by politicians and churchmen, there could be no doubt that they bore the hallmark of the political supremacy of those in power. Overseas they also signified the territorial expansion of the colonizing country, as can be seen in the station at Bombay. Something uncomfortably similar can be seen in the station at Metz in France, which in fact symbolizes the German occupation of Alsace and Lorraine.

Opposite: the station at Villepinte in the Paris region was chosen in 1977 as the target for bomb attacks by the Corsican Independence Movement, which in this way demonstrated its 'unalterable will to fight the French Colonialist State on its own ground'. (Communiqué from the National Front for the Liberation of Corsica: 26 December 1977)

In the 1830s the railway did not yet seem to have a great future. Some towns refused to admit the railway and the construction of stations. Bankers were the first to understand the importance of this new means of transport and its possibilities for the use of capital. They financed the first lines, built without any national plan in the 1830s in England, Belgium, the United States, Germany and France, and they founded the first companies. The railway gave a new impetus to the Industrial Revolution and gave rise to a spirit of enterprise which looked for profit and changed the whole social and economic pattern of each country in the process.

1
American poster of 1839. An appeal to the people of Philadelphia to prevent their town from becoming a suburb of New York. (Photo Union Pacific Railroad Museum)
2
English cartoon satirizing the rejection of the railway by the aristocratic University of Oxford. The University opposed the building of a station at Oxford, as much from fear of the intrusion of a working-class population as from a desire to protect Oxford from the railway itself.
3
American cartoon of 1882 showing the railway barons sharing out the concessions on the Island of Manhattan. The railway appears as a vital weapon of capitalism. (Photo CCI)
4, 5
In the 1950s, a swathe was cut through the heart of Brussels to make way for a railway linking two main stations. The city centre still remains gutted. The name 'Place de l'Europe' has been solemnly conferred on an area which, round the central station, has become no more than a desert of disused land, parking spaces and office buildings. In a desire to mend this breach in the city and to try to revitalize a dreary quarter, as well as to promote its repopulation by those social classes which have been largely driven from the city centre, the 'Urban Action Research Workshop', which fights for the 'democratization of development', put forward in 1976 this plan for the area of the Central Station in Brussels. (Photo AAM)

6. The Age of Bare Knuckles

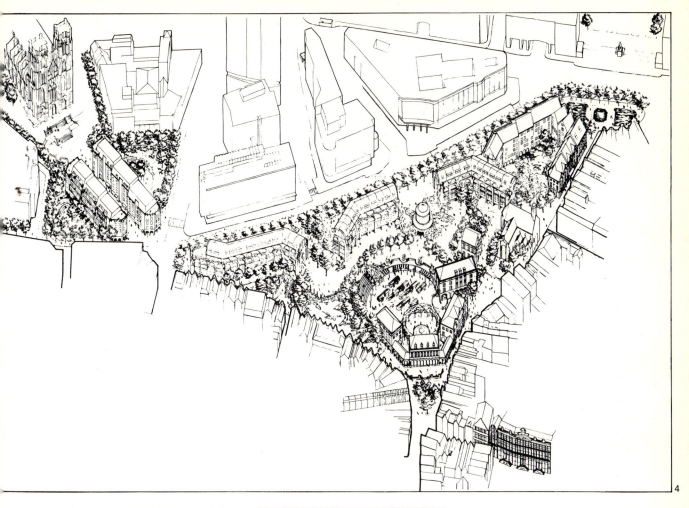

4

5

Like a lighthouse in the town, the railway station guides vast crowds, and thereby lends itself to the use of those in political power. Thus it becomes a sort of theatre in which power can display itself, as is sometimes reflected in station architecture and decoration.

1
Decoration of the façade of an Italian station in the Fascist era, a public building asserting the authority of the leader: 'Mussolini is always right'. (Photo FS)
2
Carriage of an Agit-Prop train, USSR, 1918–20. At a time of cold, famine and civil war the Bolsheviks mobilized cultural and artistic resources to win over the people's participation in the development of a new society. Propaganda trains travelled throughout the land — indeed, right up to the front lines where the Red Army was fighting the Whites and foreign troops. The trains were equipped with printing presses and projectors. (Photo Snark)

'Painters and writers will immediately take their pots and paint and, using the brushwork of their art, will illuminate and cover with designs the sides, the fronts, the hearts of towns, of stations and of the trains ever on the move.'
Mayakovsky, Decree No. 1 on the Democratization of Art, 1918.

3
Station at Springfield, Illinois, 1861. President Lincoln addresses his fellow citizens. A diorama display evokes the pioneer American electioneering tour which was later to become standard practice. (Photo USIS)

2

3

The colonial policies of the European countries resulted in two different styles: the export of Western architectural motifs and attempts at assimilating local models. These were but two faces of the same domination – white colonialism in Black Africa.

1
Station in Pretoria, South Africa, c. 1915. (Photo UIC)
2
Station of Bobo-Dioulasso, Republic of Upper Volta. (Photo Documentation Française)

Prelude to a totalitarian 'final solution': station platforms crowded with Jews, 'human cattle', destined for the extermination camps. Dreadful detail: the Nazis built a fake country station, complete with pretty flower-beds, near the entrance to Treblinka, intended to conceal from the waves of arriving victims the awful fact of the next stop, the gas chambers.

3
The rails leading to Auschwitz Camp. (Photo Jewish Contemporary Documentation Centre)
4
'The Final Destination', drawing by Willen. (In *Quiet, the Enemy is Listening*, Edition du Square, 1976, photo CCI)
5
Waiting for death on a platform. (Photo Jewish Contemporary Documentation Centre)

1

2

3

4

5

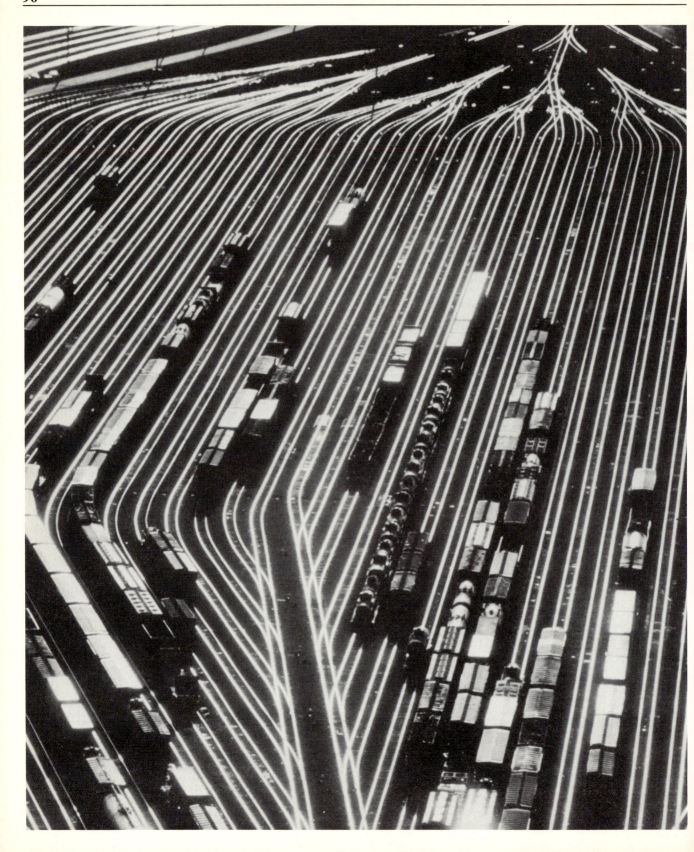

The station: unfamiliar ground

From the front, a railway station is part of the town. The square or the avenue of which it is the focus, with its hurly-burly of motorcars, its hotels with the reassuring names of towns served by the line, the urban architectural treatment — these things all provide a familiar scene for the visitor.

Behind, only fleetingly glimpsed by the traveller as the train makes its slow approach, there is the hidden face of the station, a maze of tracks, platforms and other railway structures. Outside the restricted areas of passenger circulation, only the privileged, the railway staff, may penetrate. This is the landscape of rails, criss-crossing, joining and separating, across which, in constant movement to and fro, the slow and majestic processions of trains find their way, directed by invisible masters. The rails form a sort of carpet in this strange architecture of iron and glass provided by signal boxes, gantries, water cranes, substations, etc. In this scene, the railway staff are tireless choreographers, controlling a sort of mechanical ballet which is lit day and night by the ever-changing signals holding or releasing trains.

Further on, there are engine and carriage sheds, warehouses, roundhouses, and depots — places where a staff never seen goes about its business: the engineers, the goods and parcels handlers, the maintenance men and so many others. Between the tracks one can sometimes see traces of industrial archaeology, signs of the changes in motive power, in track layout and even disused structures covered with ivy or red with rust.

Around all this, there are factories, workshops and warehouses belonging to private companies, mostly erected in the 19th century at a time when commercial activity was so largely dependent on transportation by rail. Many are now disused or no longer have their railway sidings. Large stations often have their approaches lined in this way for several miles. The backstage of a great station is a vast territory strewn with railway lines and bordered by a frieze of buildings in use or in ruins.

To be practical about this, there are here vast reserves of space and wealth, of which the possibilities have only recently come to be recognized. In many cases these are the sole remaining substantial areas in large towns that have not already been fully exploited, and for the most part they are public property. They may offer the only opportunities for major development and the sole possibilities of arresting the depopulation of town centres. Paris has already realized this, and of ten large developments outlined in a scheme approved in 1977, six are placed behind the six major Paris stations.

The largest of all these plans covers 2,000 acres around two neighbouring stations, one on each side of the Seine, the Gare de Lyon and the Gare d'Austerlitz. Around them and on their territory it is planned to create a completely new urban centre within Paris.

Opposite: marshalling yard in Kansas City, Missouri. (Photo E. Kristof, USIS)

1
Interior of an electrical power switching station. (Photo SNCF)
2
Interior of a signal box at Johannesburg, South Africa. (Photo UIC)
3
Locomotive round house of the PLM line in the Gare de Lyon, Paris, during the 1930s. (Photo SNCF)
4
The old No. 5 box at Laon, France, c. 1924. (Photo SNCF)
5
Signal gantry and signal box at the entrance to a British station. (Photo NRM)

3

4

5

1
Climbing out of the firebox of a locomotive under repair, France. (Photo La Vie de Rail)
2
The old workshops of the lampmakers at Villeneuve-Saint-Georges, Paris region, during the 1930s. (Photo SNCF)
3
Former boardroom at the Gare St-Lazare in Paris, c. 1900. (Photo Archives of Paris)

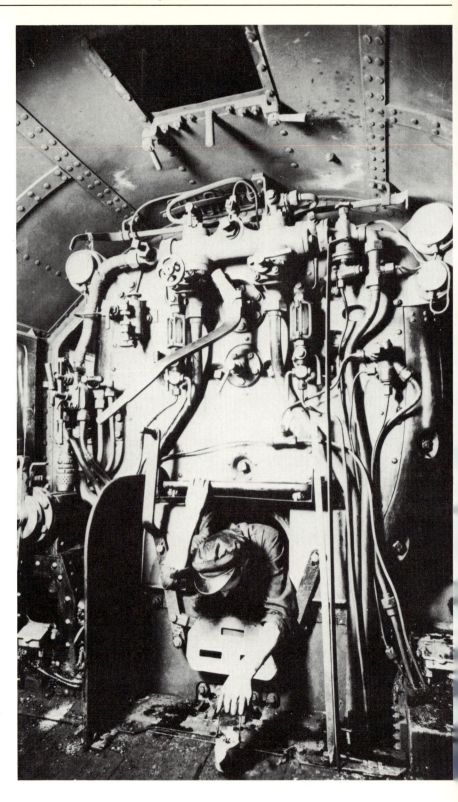

2

3

1
The great expanse of railway tracks behind
Zürich Central Station, Switzerland. A
remarkable opportunity to create above these
tracks a 'city within a city'. (Photo Comet,
Zürich)
2
'Seine-sud-est'. An immense development
plan centred on two Paris stations. The
development directives in the 'Scheme of
Guidance' of the City of Paris (approved in
1977) are very revealing: of ten major
developments which remain possible within
the central area of Paris, six are sited behind
the six largest Paris stations. The largest and
most ambitious plan is spread over 2,000
acres round the two great stations situated
on each side of the Seine, the Gare de Lyon
and the Gare d'Austerlitz. Here it is proposed
to create the Seine-sud-est scheme, a
completely new urban centre within Paris.
(Photo IGN)

STATIONS
An Endangered Species

An information program to encourage the reuse of railroad stations. Sponsored by the National Endowment for the Arts in conjunction with Educational Facilities Laboratories

An endangered species

In most of the developed countries of Europe, the period since the Second World War has seen the abandonment or the dismantling of a large part of the railway system, especially branch and local lines. In France, roughly 300 kilometers of railway track have been taken out of service every year since 1960. Recent reports to governments have envisaged further closures. By now, large numbers of towns and villages have lost their stations, their points of connection with the rest of the country.

Yet many now realize the importance of railways, as the effects of the energy crisis hit other forms of transport. Railways, too, offer the best hope of revitalizing rural areas suffering the downward spiral of depopulation.

In the USA, where Americans have almost entirely forgotten the experience of long train journeys thanks to railway closures and the spread of air travel, the government has been trying to restructure a national passenger system through Amtrak.

For the stations themselves, a similar change of attitude is now becoming apparent. In the 1960s it was felt that the great historic stations should be swept away as old-fashioned and inconvenient relics of the 19th century and be replaced by modern structures matching the new technology of the railway. Some of the stations so destroyed were of considerable historical and architectural importance, and were greatly loved by the public. Such was the fate of

Pennsylvania Station in New York, for instance, and Euston Station in London. But 19th-century architecture, so recently despised or misunderstood, is now appreciated and restored to favour. In Zürich, the Central Station was, a few years ago, to be demolished and converted into a vast 'urban renewal' development. This plan has now been abandoned and the station is being restored to its original condition without endangering its effectiveness for railway purposes.

The public is now more watchful. It readily mobilizes to oppose the demolition of places which seem indispensable to the town for aesthetic, symbolic or sentimental reasons. Grand Central Station in New York, threatened with demolition, stimulated a great popular campaign which ended up before the Supreme Court in Washington. In Great Britain, a similar campaign arose in response to proposals to demolish Liverpool Street Station in London, and a recently founded organization, SAVE, has undertaken to arouse public opinion on behalf of the heritage of railway architecture. It has forced British Rail to adopt a more responsible attitude to the buildings in its ownership. The time seems at last to have come for a civilized reconciliation between the need to cherish the better examples of railway architecture and the need to modernize the operations of the railway. The recent examples of the stations at Zürich, Copenhagen or Paris Nord can serve as examples.

Opposite: this American poster, published in 1977 to draw public attention to the dangers threatening railway stations, was part of a wide-ranging campaign which is still going on in the USA to preserve this inheritance and to find ways of giving it new meaning and vitality. (Illustration by Michael Goldberg, CCI)

1, 2
The case of Euston Station in London is very
revealing of the spirit in which
'modernization' of railway architecture was
approached in the 1960s. Above, the great
hall built in 1849 and demolished in 1962.
Below, the concourse of the new Euston.
(Photos NRM and BR)
3
Since the 1960s the demolition of large
railway stations has become a frequent
occurrence in the USA, sometimes making
front page news. Here is the end of the
station at Portland, Maine, built in 1868 and
demolished in 1961 to make way for a
shopping centre. (Photo CCI)
4
The evolution of the British railway network:
(left) its origins in 1840, (centre) at its
height in 1925, (right) as predicted for 1980.
(Photos SAVE)

Turn On Your Lights In The Daytime This Holiday Weekend To Remind Others To Drive Safely

The Weather:
Official Bureau Forecast
Fair, Warm, Humid
Today and Saturday
(Complete Report On Page 2)

Portland Press Herald

Our Number Is:
SPruce 5-5811

★ ★ ★ ★ 26 Pages

VOL. 100—NO. 61   Established June 23, 1862   PORTLAND, MAINE, FRIDAY MORNING, SEPTEMBER 1, 1961   Second Class Postage Paid At Portland, Maine   PRICE SEVEN CENTS

Union Station Tower Comes Tumbling Down

An historic Portland landmark passed from sight in a matter of seconds Thursday afternoon. While hundreds of persons watched, workmen razing Union Station toppled the clock tower with a swinging steel ball.

Last to leave the tower was a seagull (inset, left photo). Nicknamed "Willie the Hermit" by workmen, the bird has been perched on top of the tower daily since demolition started six weeks ago.

The station, built in 1888, is being razed to make way for a shopping center. (By Staff Photographers James, Roberts, Morrison and Johnson)

World Shock Follows Red N-Test News

By THE ASSOCIATED PRESS

Moscow's decision to resume nuclear testing provoked worldwide shock waves Thursday, even among some nations that often lean toward the Soviet line in the cold war.

peatedly asserted that the nuclear armaments race is the greatest danger facing mankind.

The Indian prime minister feels a disarmament agreement is necessary which would re-

Massive Bang Expected Soon

President Holds Back On New Nuclear Tests

WASHINGTON (AP) — President Kennedy declared full confidence in U.S. atomic might Thursday and held back on resumption of nuclear weapons testing while the Soviet go-ahead decision soaked in on a shocked world.

Kennedy denounced the Soviet action as "atomic blackmail."

cided at least for the time being against resuming atomic testing.

"That's the way I would interpret it," Hatcher replied, "but it's up to you."

An indication of the administration's current trend of thinking and reports from Belgrade showed some negative reaction already was setting in against the Soviet Union.

tion was keeping an eye on the effect the Soviet move was having on a conference of professedly neutralist states starting Friday in Yugoslavia.

Many of these unaligned nations are opposed to nuclear testing.

This business of the

tol Hill there was angry talk aplenty.

Sen. Thomas J. Dodd, D-Conn., who drafted the resume-testing-now resolution, said the U.S. suspension of testing was "the most fatuous blunder in our history."

Sen. Thomas H. Kuchel, R-Calif., told the Senate, "The sham and hypocrisy of the Soviet Union stands out in bold relief be-

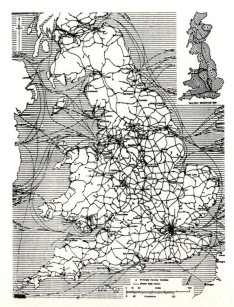

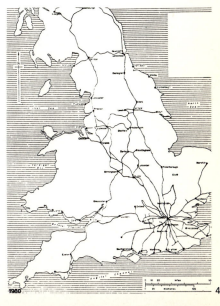

1848   1980

3

4

In most Western countries the greater part of our railway architectural heritage derives from the 19th century. The wonder that these buildings excited at the time was replaced around the 1920s – while functionalism was in the ascendancy – by contempt or indifference. In such a climate of opinion, disfigurement of stations became rife, with the building of ugly extensions or clumsy attempts at modernization.

1
Great hall of Zürich Central Station; a fortunate exception. Here restoration is under way, and it will soon be cleared of its unsightly accretions. (Photo ONST)
2
Minden Station, Germany. An example, now current in Europe, of an unhappy attempt to impose 'contemporary taste' on an old railway building. (Photo DB)

In a number of Anglo-Saxon countries, the demolition of stations has led to the mobilization of public opinion in an attempt to oppose these destructive actions.

3, 4
Station in ruins and publicity flyer calling New Yorkers to a mass demonstration outside the Supreme Court to safeguard Grand Central Station. (Photo ACL)

3

New uses for an old asset

In thirteen years, between 1963 and 1976, British Railways closed numerous lines and 3,539 stations fell into disuse. In the USA, of 40,000 stations built in the 19th century, only about half remain, and most of these are no longer used. In all Western countries this sort of thing is happening, partly for economic reasons (especially where, as a consequence of nationalization, there is no need to maintain competitive stations in the same area), and sometimes as a result of changes in the siting of industries, their decline, and associated population movements. As a result, both town and country have many railway installations, stations, warehouses, engine sheds, etc., which are no longer in use.

Not many years ago it would have been normal for any structure of this kind occupying a valuable site to be demolished forthwith to make way for some new property development. But times have changed and there is a new awareness of possible alternatives. In recent years, 19th-century architecture in general has ceased to be despised and has come to be appreciated, and even more recently the quality of much industrial architecture has come to be recognized as worthy of preservation. Perhaps the very rapidity of change in our urban landscape has made us more appreciative of some of the nobler relics of even a century ago. To this has been added, very recently, what might be called an ecological conscience which has produced a welcome reluctance to waste resources. Why demolish these substantial and splendid old buildings? The care and craft that went into their creation is not likely to be repeated in the making of their successors.

Thus redundant stations and other railway structures no longer in use are beginning to be regarded as capable of fulfilling some new social purpose. Their previous use has frequently provided them with excellent connections within their towns. They are adapted to receive large numbers of people. Stations can well become museums, markets, sporting, cultural or commercial centres, theatres, restaurants or other places where people gather. The surrounding land may also provide parks, botanic or zoological gardens. There have already been a few brave pioneering projects and some have achieved remarkable success.

Opposite: built in 1854, this old station in Strasbourg was one of the first to be converted to a new use. In 1972 it became a public market. (Photo Pilloux, VDR)

1
A British rural station, no longer used, converted into a country residence. (Photo NRM)
2
The American rural station at Winton Place (1879) was dismantled and transferred in 1969 to the open-air museum of Shaton Woods Village. (Photo Miami Purchase Association)
3
Grand Union Station in St Louis, Missouri. Built between 1891 and 1894 to the design of Theodore Link, this gigantic station is now redundant. A scheme is being prepared for its conversion into a commercial and cultural centre. (Photo Library of Congress, Washington)
4
The railway works at Crewe, built in 1840 and subsequently enlarged; demolished in 1973. (Photo National Monuments Record, London)

1

2

3

4

1
The old locomotive round house at Chalk
Farm in north London, built to the designs
of Robert Stephenson in 1847. Since
conversion in 1962, the building has found
fame as 'The Roundhouse' theatre. (Photo
NRM)
2
Project for converting the old station in
Pittsburgh, Pennsylvania, into a large
commercial, office and leisure centre. (Photo
Pittsburgh History and Landmark
Foundation)
3, 4, 5
The station in Brunswick, West Germany,
was opened to traffic in 1845. Its design, by
Carl Theodor Ottmer, was long considered to
be one of the finest in the country. It was
damaged during the Second World War and
offered for sale in 1960. A brilliant design by
the architect H. Westermann has converted it
into the local headquarters of a bank.
(Photos Westermann)

1

2

3

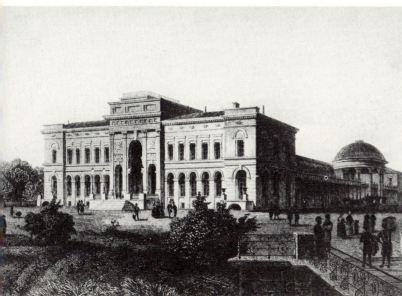

4

5

1
The Gare d'Orsay, built in the centre of Paris
by the architect Victor Laloux between 1897
and 1900, combined a sumptuous hotel with
a lower-level railway station. Both fell out of
service several years ago. On the initiative of
the French President, this splendid specimen
of railway architecture is to be transformed
into a museum of the 19th century,
complementing the Louvre (on the opposite
bank of the Seine) and the Centre Georges
Pompidou nearby. The conversion scheme,
devised in 1973 by Patrick O'Byrne and
Claude Pequet, is by far the most ambitious
and prestigious station rehabilitation project
in France. The net usable space will be
43,000 square metres, and it will require a
staff of 700. It is likely to open, at the
earliest, in 1983. For several years before the
work began, the station was known as the
'Théâtre d'Orsay', home of the Renaud-
Barrault Theatre Company; part of the
structure also housed the Drouot-Rive
Gauche auction rooms. (Photo CCI)
2
A Swedish scheme for using abandoned
railway lines and stations for cultural
activities. (Photo CCI)
3
The station of the Provence Regional
Railway at Nice was built in 1892 and has
been threatened with demolition for some
years. The architects Philippe Robert and
Bernard Reichen submitted to the Mayor, in
1978, a plan for the partial conversion of this
station to allow it to become the civic centre
of its district, while preserving its railway
function. (Photo CCI)
4
Advertisement which appeared in the French
business press in March 1978, reflecting the
paradox of a country which aspires to export
its know-how for the regeneration of disused
railway lines abroad, while numerous lines in
France itself remain abandoned. (Photo CCI)
5
Italian scheme for reusing railway equipment
to make a hotel and restaurant, 1930;
architect, Portaluppi, 1930. (Photo P.
Saporito)

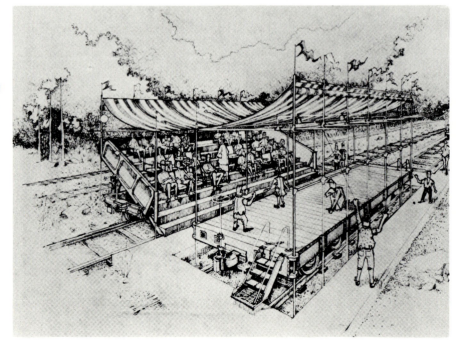

7   3

Au bout de
ces voies
la fortune d´un pays.

De nombreux pays neufs ont compris qu'ils ne connaîtraient pas d'essor réel sans un bon réseau ferroviaire : sur tous les continents et sous tous les régimes, le rail demeure l'outil de base du développement économique.

Il s'agit donc aujourd'hui, dans des pays de plus en plus nombreux, de refondre plus ou moins complètement des réseaux fatigués. Et cela, c'est la vocation même de Francorail-MTE.

Équipements mobiles ou statiques, formation, construction, Francorail-MTE propose aux pays neufs de mettre entre les mains des techniciens na-

tionaux l'outil ferroviaire moderne nécessaire à leur développement.

Poursuivie depuis 10 ans – Turquie, Brésil, Guinée, Cameroun – cette politique de coopération technique vaut aujourd'hui à Francorail-MTE une place de leader mondial dans le domaine ferroviaire.

Cela se sait partout dans le monde. Et cela commence à se savoir en France.

Francorail-MTE, GIE régi par l'ordonnance du 23.9.1967 regroupe Carel-Fouché-Languepin, Creusot-Loire, De Diétrich, Jeumont-Schneider et MTE, 2, rue de Léningrad 75008 Paris.

francorail mte

Francorail MTE exporte ses techniques
et son savoir-faire.

4

5

In search of an image

Recent stations have often been the weakest link in projecting the image of the modern railway. Since the Second World War, railway architecture has been in a decline and the station seems to have lost its identity. In many countries the elimination of competition as a result of nationalization or amalgamations has ended the architectural rivalry that once gave stations a character related to their railway and its territory.

The grandiose style — still common enough between the Wars — seems less justified as the railway loses its pre-eminence. At the same time, the spread of an international style of architecture, rejecting all historical and cultural associations, has swept away local design traditions. The exuberance and eccentricity of ornament, so popular in the past, is also banished. Anonymity is the result. It may soon be impossible to distinguish at first glance between a railway station, a power station, a government office, or an airport.

This is even happening in the Third World, where the railway has sometimes come to be regarded as a symbol of colonial times. All the same, the modern stations in these countries are no less Western in inspiration, because of the enthusiasm of Third World countries for modernization.

The new station at Berne in Switzerland is modelled on a shopping centre: it has even been necessary to write the word 'Station' three times across the frontage, so characterless is the building. Inside, the space is organized on three levels, connected by sets of escalators and full of shops. It is almost by accident, turning a corner in the basement, that one discovers the trains.

At Argenteuil in the Paris suburbs, the station looks like the place where one pays the tolls on a motorway. At Edmonton in Canada, at the new Central Station in Utrecht, or at Maine-Montparnasse in Paris, the first impression (and a sombre one at that) is of an office building.

There are also many stations, like that at Ottawa in Canada, which copy the style of an airport: this borrowing from the architecture developed by a rival, together with many imitations in detail, suggests a massive inferiority complex.

The station is in search of an image. Until it finds one, its characterless modern manifestation will continue to be yet another contribution to the dreariness and inhumanity of modern town centres.

Opposite: Louvain la Neuve Station in Belgium, 1976; architect, Yves Lepère. Having for some twenty years been deprived of all forms of monumental and symbolic expression appropriate to the city, modern stations are now often closely integrated into multi-purpose complexes where offices and shops, cultural and educational activities exist side by side. Sometimes this combination of functions enables the station to resume a meaningful place in the centre of cities, as has been the case here. (Photo Némerlin)

Railway stations often sought to emphasize the particular character of the towns or the districts to which they belonged. Their architecture was unmistakably 'railway'. This is now a thing of the past. Not only do modern stations give little outward sign of their special role, but their lack of character fails to associate them with their geographical location: they could be anywhere in the world.

1
The station in Ostrava, Czechoslovakia, 1966 to 1974; architects, Lubov, Lacina and Vlasta Dorisa. (Photo V. Slapeta)
2
A new station in India, 1969. (Photo X)
3
Eindhoven Station, Netherlands, 1956; architect, Van der Gaast. (Photo NS)
4
Kursk Station, Moscow. (Photo Tass Agency)
5
Warsaw Station, 1975. (Photo Interpress)

'The best stations are those one does not notice, which are so readily accepted by the public that they are simply a part of everyday life.'
R. Humbertjean, former Chief Architect of the SNCF, in 1973.

3

4

5

Once proud to stamp their own identity on each town, stations now copy the dreary style of other business buildings.

1
Argenteuil Station, Paris suburbs, 1970; architects, R. and R. Dubrulle. This station resembles the toll barrier of a motorway. (Photo R. Dubrulle)

2
The old and the new Central Station in The Hague, 1972. (Photo Municipal Archives of The Hague)

3
Central Station in Berne, Switzerland, 1965–74; architects, J.-W. Huber and P. Bridel. A distracting resemblance to a shopping centre. (Photo CFF)

4
New uniform for an SNCF hostess inspired by those of the airlines. (Photo SNCF)

5
Edmonton Station, Canada. More like an office building than a station. (Photo Canadian National)

6
The station in Ottawa, Canada, 1958; architects, J. B. Parkin and Associates. The model is clearly that of an international airport. (Photo Canadian National)

1

2

3

4

5

6

In spite of the characterlessness of
most modern stations, there have been
some signs of the stirring of new ideas.
Some architects have tackled the
spatial and monumental potential of
stations with imagination. But as yet
few of their attempts have got beyond
the drawing board.

1
The station at Roissy Airport, north-east of
Paris, 1976; architect, P. Andreu.
(Photo SNCF)
2
Station in Kungsanger in the Stockholm
area. (Photo Swedish Railways)
3
Competition design for the new Lucerne
Station, 1976; architects, Werner Kreis and
Ulrich Schaad. (Illustration Kreis and
Schaad)

3

The station in art

For 150 years, the station has served as a powerful symbol in art, although its meaning has changed. Seen originally as an object, a place, something outside ourselves, the station is now felt as an experience. This is reflected in art.

When it appeared at the beginning of the industrial era, the station was the symbol of social progress. To the bourgeoisie, conscious of their power, it was the mirror in which they were glorified through a proliferation of triumphal images. At the same time, and in contrast to this notion, there appeared a new vision and a new approach to painting. Claude Monet saw nothing symbolic in the station, but rather saw it as an opportunity to capture the impression of steam.

Nevertheless, up to the period between the two World Wars, the station inspired mainly conservative artistic expressions, often anecdotal or ideological in theme. In the First World War, the station was pressed into artistic service as a setting for patriotic tableaux. It had not lost its magical, symbolic power, but this was now limited to the context of its familiar, practical use. At about the same time, Futurist painters celebrated the station as a place of transit, a world of dynamic sensations, a symbol of modern cities.

In contrast with these dislocated visions, evocative of the violence of war, the spaces of Giorgio de Chirico, with their dreamlike apparitions, show for the first time a complete internalization of the subject: the station as experience. The station becomes a place of solitude, silence, timelessness. Behind its arcades and its piazzas, extend 'those distant horizons full of adventure' which the Italian artist described in his book *Hebdomeros.*

In the work of Paul Delvaux, the station has become intensely personal, almost like a memory from the past. It lends its particular atmosphere to the projection of the painter's fantasies. Here it is a desert-like place, an image of a dead world from which the crowds have been erased, in which individuals are alienated, alone in an empty space.

Although now inevitably treated as the painter's personal vision, and no longer as strict representation, the station can still speak as a symbol to a modern audience. However, because its status has changed in Western society, it is less and less used in painting. No longer a symbol of progress, the station has lost its monumentality; it has become a means of access to the banal and the functional.

Opposite: Fritz Gerlach, *The Station,* 1965, oil on panel. (Photo Städtische Kunsthalle Recklinghausen)

Monet, disregarding all anecdotal
details, strove to paint light effects in
the steam from the trains in the Gare
St-Lazare, whereas Karl Karger
presented the Viennese bourgeoisie
with painstaking realism, architectural
precision and decorum.

   Another artist, Fritz Gerlach, uses a
classical perspective technique in *The
Station* (p. 120). His painting sets aside
the detail of place and period in order
to suggest a weird climate of silence.

Karl Karger, *Arrival of a train at the North
Station, Vienna*, 1875; oil on canvas. (Photo
Österreichische Galerie, Vienna)

1
Fernand Léger, *The Station*, 1923. Oil on canvas. Station architecture was Léger's inspiration for an organization of lines and coloured planes, whose formal beauty declares its independence of descriptive and sentimental values. (Photo Galerie Krugier, Geneva)
2
Camille Pissarro, *Lordship Lane Station, Upper Norwood*, 1871. Oil on canvas. In this placid scene, Pissarro seems to have absorbed the traditionally British affection for railways. (Photo Courtauld Institute Galleries)
3
Giorgio de Chirico, *The Anguished Journey*, 1913. Oil on canvas. (Photo The Museum of Modern Art, New York; Lillie P. Bliss donation)
4
Paul Delvaux, *Night Train*, 1947. Oil on wood. (Photo Secrétariat d'Etat à la Culture française, Brussels)

3

4

1
Fabio Rieti, *Wall in Dijon*, 1973. Oil on
canvas. The station demystified . . . (Photo
Jacqueline Hyde)
2
Jean Le Gac, photo of the station at
Cauterets, France. Here the photo is intended
to evoke a station in a Western film.

Microcosm of an alienated society . . .

Leonardo Cremonini, *Departure*, 1972–73.
Oil on canvas. (Photo Jacqueline Hyde)
4
Eduardo Arroyo, *Frankfurt-am-Main Station*,
1970. Oil on canvas. (Photo Giancarlo
Baghetti)

3

4

Stimulus to the imagination

Image of a new age, the station presented itself to our forbears' fascinated gaze as an imposing monument concealing noisy and mysterious machines: steam locomotives. But with the passage of time, familiarity has robbed the station of its glamour, and now people hardly even look at it.

How did popular culture absorb this newcomer which so transformed the daily scene? At first it was charged with an aura of invention and novelty. The opening of a station was a festive occasion, worthy of being memorialized not only in paintings and engravings but in other ways as well. The things used for this purpose — plates, tea sets, calendars, etc. — emphasize the role of the railway station as a symbol of technological civilization, a thing to be honoured in middle-class homes. Its strangeness — the sense of the station being an intruder — was soon overcome. Caricatures, jokes, and other manifestations of popular culture effectively tamed this difficult animal.

The station was also the open doorway to travel. Travel, seen at first as perilous, lost its terrors as it became more common. 'I've just arrived at Amiens and send you greetings': such, inscribed on a postcard, became the watchword of the modern traveller.

From the beginning of the 20th century the media reflected the widening field of possibilities offered to more and more people by the new means of transport. The cinema was not slow to catch on. The station was often presented as a theatre of life, a place where destiny could make its impact in the most unexpected ways. In Westerns and in war films, the station became a place of danger through which people were obliged to pass. Alfred Hitchcock made a convention of the station as a place of suspense in detective films, and others copied him until it became a cliché.

The station and its mythology lend themselves splendidly to the expression of dreams and fantasies: the station as a woman, the station as a bottle, the station in a bubble, the centre of the world for Salvador Dali. Toy railway stations continue to enchant children. It's the home of the train, and the railway is always the most splendid of toys. It has long been like that, though they don't make toy stations as they used to do, and the old ones made of tinplate are now desirable collector's pieces.

The station and the ways in which it has been illustrated are no longer symbols of a new world, but they still evoke in us a certain nostalgia for the never-never land of yesteryear.

Opposite: a combined locomotive–station, drawing by Rowland Emett from his book *The Early Morning Milk Train*, 1935. (Photo Planchet, CCI)

The station was strongly surrounded by all the symbolism of travel. Place of adventure, experienced or imagined, its image was projected towards a fabulous future in the universe of science fiction, where it still exercises its evocative power in people's imagination.

1
Postcard: example of a series, very popular during the 1920s, in which a short expression of politeness accompanied the name of the station of departure or arrival. (Coll. Kneebone)
2
Wood-engraving of one of the first stations in Holland: evidence of the popular craze for railway stations. (Photo Municipal Archives of Amsterdam)
3
The Railway from Paris to the Moon, lithograph published in *La Mode*, December 1839. (Doc. Bibl. Nationale, photo Planchet, CCI)
4
Futuristic drawing by Biedermann in 1916.

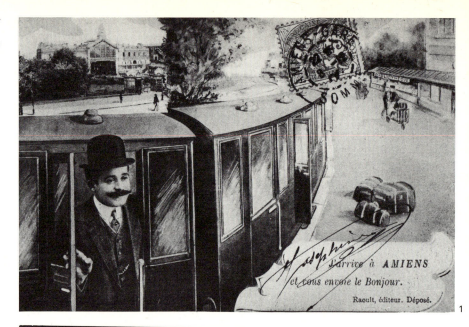

1

2

PREDICTIONS

2

L'industrie ne connait plus d'obstacles.— L An 40 voit se réaliser tous les projets en l'an — Chemin de fer suspendu, de Paris a la lune—Multitude de trous faits a cet astre par un grand nombre de Banquiers, Gerants, Directeurs, Administrateurs et autres jeunes gens sans experience.— Horrible secheresse qui detruit tous les actionnaires.— Un savant horticulteur découvre que la graine de cornichons les fait repousser—Te Deum chante a cette occasion.

Pronostic.   L'Actionnaire et le cornichon
Dans tous les temps se mangeront.

3

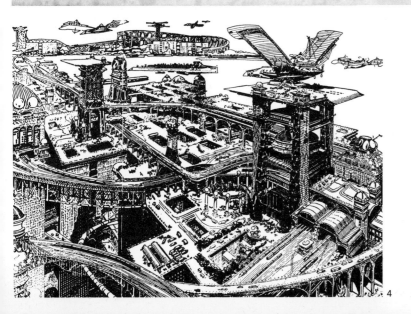

4

For a long time, the station was seen in popular art as a foreign and ill-assimilated body that housed an infernal invention, the train. A railway catastrophe, freakish weather that gave a station a surprising appearance – these were immediately seized upon by the media. Travellers numb with cold and fear at the start of a train journey were the easy prey of caricaturists. In our own time, Saul Steinberg has – with great economy of means – given us a poetic and synthetic image of the station.

1
Saul Steinberg, drawing of station. (From *Saul Steinberg*, Gallimard, Paris, 1956)
2
Tintin in the Soviet Union: the hero's triumphal return to the Brussels Gare du Nord. (Hergé, Editions Castermann, 1929)
3, 4
A station in the extreme north of Sweden encrusted with ice, photographed during a particularly severe winter on 15 January 1929. (Photo Nordiska Museet, Stockholm)
5
People (and a dog) muffled up in bags to protect themselves against cold and accidents during a journey; anonymous French caricature, end 19th century. (Photo Planchet, CCI)
6
Third-class passengers frozen stiff, Honoré Daumier, lithograph *c.* 1870. The luxury of heating, if it was provided at all on trains of the time, was usually reserved for those who went First Class. (Photo VDR)
7
Railway accident at Charenton Station: one of the fears of 19th-century travellers. (Photo Planchet, CCI)

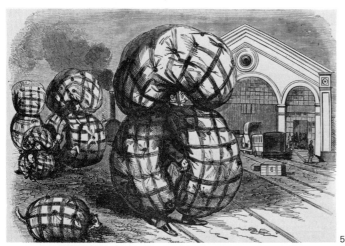

For the sake of entertainment, this superhuman invention was domesticated and reduced in size. The railway network, with its constellation of stations, amused children and adults alike. The point of the game – man dominating the machine – sometimes took on a piquant aspect. Thus Gaston Menier turned his dining table into a station for toy trains carrying plates of food.

The station became for caricaturists and illustrators a kind of observation post from which they could draw the behaviour of people.

1
An absurd ceremony at the opening of a British country station, drawing by W. Heath Robinson, 1935.
2
The last batch back from holiday, caricature by Alain Saint Ogan, 1936.
3
Choo Choo, the story of a little locomotive who goes in search of adventure, drawing by Virginia Lee Burton, 1944. (Faber and Faber, London)
4
Gaston Menier's dining-table with its little electrical railroad, in *La Nature*, 1887.
5
Illustration from a Hornby catalogue of railway models, *c.* 1947. These children's toys have since become collectors' items.

3

4

2

A. Levy. A century of models trains. Coupar Books 1975. p 140

Postscript

Dividing point – or transition point – between two modes of transport, two rhythms, two landscapes, between a very recent past and an immediate future, the station as experienced by its users can, without too great a stretch of the imagination, summon up the more or less fleeting moment when the traveller waits for his train, meets it or leaves it. The duration of this moment defines a 'time' in the same sense as 'time to think' comes before a decision, and 'time sheet' structures a school day, and 'May time' and 'holiday time' return every year, with infinitely various emotional impact.

In our culture, always on the move, time at stations may be seen as an attempt to stop the motion, so as to take a series of snapshots or short movies showing a cross-section of the lives of individuals, singly or in groups, some of whom experience the moment vividly, while others attach no importance to it. There is an abundant harvest: joys and reunions, farewells and the pangs of parting. People absorbed in the emotions rub shoulders and take no account of one another. They mix with travellers on pilgrimage or en route to sports events, people on business trips or setting off for holidays in the sun, young people leaving home or night-shift workers passing through in the early morning. Time at stations includes departures for the Front in 1914–18, waiting for the last refugee train at the Gare d'Austerlitz in June 1940, the last journey of those deported to concentration camps in 1943–45.

The stations of Bombay and Calcutta by night, or the encampment of immigrant workers in the station at Frankfurt, remind us that the French word 'gare' (station) is derived from the Germanic 'wardare': to guard. So time at stations becomes the time of shelter, of refuge, for those who are afraid or do not know what tomorrow may bring.

A century ago, the railway brought about unified time zones within national frontiers. Today, the station clock presides over the square and sets the pace for those approaching the town centre or the train. Its precision and its inexorability give it power to regulate the hours kept by millions of people, for the train cannot be kept waiting and does not respond to imagination or dreams. Thus, the time of day is crucial in railway stations, and each user accords it a significance of his own.

For many decades, the station was a special place because of its size, its architecture and decoration, its significance as an urban symbol. There are those who think that the railway station belongs to the past and that one can refer to 'the station age'. The last 150 years have indeed seen an evolution in the appearance of stations, just as churches and telephone headquarters have changed in order to satisfy new needs or to employ new technologies. In the 19th century the station was in the forefront of modern development; it was a city's showcase of high technology, a symbol of urban growth, a statement that the city had its place in a national network. The station had a self-confident façade and huge vaults of cast iron or glass, which were a necessity in the age of steam locomotives. In our own day, architecture of monumental or gigantic scale is still used for 'modern' buildings, but those modern buildings are now airports, television towers, nuclear power stations, sports stadiums, cultural centres, etc. These establishments will also have their day, as new technologies – particularly in energy production and telecommunications – are developed.

The railway station has been the first to return to the ranks. Diesel or electric locomotives allow roofs to be lower, while the greater frequency and shorter length of trains mean that platforms can be shorter too. Considerations of passenger convenience, and the economic advantage of using both ground and underground levels, lead to a link between railways and urban transport networks. This gives us stations below ground level while at the same time new conurbations bring new stops and new stations into being.

At a moment when many countries are completing the process of modernizing rolling-stock, relaying tracks and even extending them, when thinking about the relationships between air, road and rail transport has made it possible to redefine the particular role of the railways, a new 'age of the station' can be said to have begun.

This development affects too many people for it to be left entirely in the hands, however expert, of bureaucrats and engineers. The railway station deserves closer public scrutiny of its origins, or its emotional impact and of the changes which will affect its future.

Jacques Mullender
Director,
Centre de Création Industrielle,
Centre Georges Pompidou, Paris